Table of contents

Foreword

"Chemical disaster in Bhopal", "Pesticides threaten water supply", "Rhine-water unsuitable for drinkwater after pollution with pesticides", "Toxic chemicals found in residential area", "Farmers addicted to pesticides". Just some of the newspaper headlines which nowadays confront the Western public with such regularity that they are gradually losing their power to shock. While the rise in media coverage has certainly contributed to the enormous increase in public awareness of the problems facing agriculture, it cannot be said that this has yet led to any significant changes. Most European countries have indeed banned some pesticides known to be highly toxic or very persistent. However until now, the use of pesticides in for instance the Netherlands has increased year by year. And unfortunately, the situation in the Netherlands is by no means exceptional, but indeed common practice.
In the Third World also, the problems associated with the use of pesticides can be ignored no longer. Development workers daily involved in rural development are confronted regularly with questions and problems on the subject of crop protection.
However, as the main source of advice is provided by the pesticide manufacturers who naturally tend to promote their own products, reliable, unbiased information is hard to come by. To rectify this situation, it is essential that development workers have access to alternative sources of information about crop protection. Practical literature which is also accessible to those who lack expert knowledge of agriculture, is difficult to find.

To alleviate this situation to some extent, a pesticide working group was set up within CON, the Dutch Centre for Development Work. CON is an association of discerning development workers whose objective is to support their colleagues in the Third World through positioning their work in a broad socio-economic and political framework. Furthermore, CON uses the experience of development workers to raise awareness of the Third World problems in the Netherlands. At the home-base in Wageningen, approximately 40 volunteers and two part-timers work to achieve these objectives. CON publishes two journals: "Contekst" and "Wanawake", beyond which they also conduct research into specific themes such as the problems in the field of crop protection.
This research resulted in 1984 in the publication of "Kleine boeren en gewasbescherming in de Derde Wereld" (Small-scale farmers and crop protection in the Third World). The intention of this book was to reduce the use of pesticides in the Third World by drawing attention to their negative effects. It also stressed the worsening socio-economic position of the small-scale farmer in the Third World caused by the introduction of western capital-intensive technologies, such as chemical crop protection. The over-riding theme was in fact that new crop protection technologies should be better adapted to the local situation in each of the countries concerned.
The book was a success and quickly sold out, but some criticism also arose. In particular, it was

considered that too little attention had been paid to the practical application of these "adapted" crop protection methods. These criticisms led to the decision not to reprint the book in its original form, but to prepare a new publication which would deal extensively with the practical aspects of environmentally sound methods of crop-protection.

Naturally, this re-worked version does not pretend to provide immediately applicable solutions to the problems in small-scale agriculture, but an attempt has been made to provide the groundwork for possible working-methods.

In addition to the above amendments, we felt it worthwhile to give the new edition a wider area of distribution, and to this end, decided to have the new version translated into English. More thought has also been given to a clearly structured lay-out.

To make possible publication of a book like this, we looked for likely sponsors and a publisher. CTA (Technical Centre for Agriculture and Rural Cooperation) was found willing to subsidize the translation. Other contributions were made by the Netherlands Ministry of Education, the European Year of the Environment and the Algemeen Diakonaal Bureau (General Office of the Churches). TOOL agreed to undertake publication of the book. The institutions who gave subsidies are greatly thanked for this, without their help this publication would not have been possible. For more information about CTA and TOOL see appendix III.

Many people have compiled and written material for this book, and a list of all contributors can be found at the end of the book. The editors would like to extend their warm thanks for the endeavours of all those who have been involved in bringing this book into being.

We hope that this publication will prove valuable to development workers in the Third World as a useful supplementary source of information. We realize, of course, that the subject of agriculture in the Third World is once again being approached from a western point of view, but unfortunately, we were not in a position to discuss the selected methods with the farmers themselves, and therefore have had to leave this to the 'workers in the field'. Reactions from development workers and extension officers about the practical value of this work would thus be warmly welcomed.

The editors,
Wageningen, September 1989

Part I
Outline and introduction to problems in crop protection

Part I of this book contains a bird's-eye view of the problems associated with crop protection in the Third World; it examines the consequences of the transition from traditional farming to an agriculture which makes use of modern techniques (e.g. chemical control), and also reviews the historical development of crop protection.

1. Introduction

1.1 Farmers, agriculture and pesticides in the Third World

Agriculture in the Third World is coming under increasing influence of modern technological developments, which mostly originate in the Western World. These techniques are in pursuit of maximum physical production levels. The economic interest (production objective) dominates other social interests, such as conservation of nature and the environment, sustainable production, employment, a fair distribution of goods, a democratic rule, etc.
Roughly described, Western technology is committed to reducing labour (labour is replaced by capital (machines) and by energy (fuel)), to urbanization, to a good supply of skilled labour and to an industry-oriented market. Not surprisingly, Third World import of such technologies, which are minimally if at all adapted to conditions prevailing there, does not proceed without problems.
One of the sectors in which these problems are clearly visible, and even affect the agricultural community itself, is that of crop protection. There is a great amount of pressure on Third World agriculture to increase productivity to provide for its still increasing population. Pests cause enormous losses. The application of chemical control technologies which proved so effective in the Western World could not be more logical. This was the reasoning of the Green Revolution strategies prevalent in the sixties. Since then, chemical products (pesticides and fertilizers) have made a firm entry into small scale agriculture in the Third World.
In the meantime the Western World has become increasingly concerned about the excessive use of chemical products in agriculture. Almost daily, the media report on negative consequences of the highly intensive production methods. There are now few places where pesticides cannot be found in the environment: chemical industries dump waste from pesticide production in rivers; the process of spraying pesticides has caused contamination of drinking water; chemical residue is left on most agricultural and horticultural produce; sensitivity of some crops to herbicides applied in previous seasons has halted their further cultivation, and so on. Because of this, more and more thought is being given to a restricted use of pesticides. The negative effects on nature and environment are then reduced, as well as the destruction of the production environment, so that sustainable agriculture becomes possible. Moreover, it is also important that the hazards to public health are reduced. In 1983 there were 2 million cases of poisoning by pesticides; 40,000 of which were fatal. As the use of pesticides has increased even more since then, it can be justifiably assumed that the number of victims has also risen. The risks to human lives and to the environment are now so great that there is no longer any question about the necessity for changing to crop protection techniques which are far less reliant on chemicals.
Information on the negative aspects of pesticides in the Third World is still scarce, as too are funds for conducting research into the effects of chemical pest control on public health and

Figure 1.1. Pesticides are applied almost everywhere in the world nowadays, also by people who are not trained, especially in Third World countries.

11

Figure 1.2. A Third World pesticide user. In the Third world applicators of pesticides are advised only by a few people: by the pesticides middleman, sometimes by an extension officer. (left)

Figure 1.3. A Western pesticide user. In the West a lot of services are concerned with the safe and efficient use of pesticides, negative consequences therefore are discovered and tackled sooner or later. (right)

environment. This book is an attempt to reduce this lack of information to some extent. One important objective in its writing is to provide development workers with a starting point for achieving a sustainable and environmentally sound agricultural system.

Another major point is concern for the social position of small-scale farmers in the Third World. The agricultural sector is of extreme importance as a considerable source of income for both national governments and the rural population. It also provides a significant market for the chemical industry. The danger that small-scale farmers will become increasingly dependent on pesticide manufacturers should not be underestimated. Another source of concern is the ever-increasing concentration of agro-chemical industries and the take-over of seed-breeding companies by these multinationals. A monopoly of the agricultural inputs market would enable multinationals to decisively influence farming techniques, in the West as well as the Third world.

The widely spread use of pesticides is partly due to the advantages they can offer. Pesticides are effective and reliable, their use can prevent loss of yield and reduce risks for losses. They work quickly, which makes them suitable for use in emergency situations, and frequently they are the only remedy when crops are under immediate threat of infestation. These advantages have led to pesticides being used throughout the world.

On the other hand, the use of pesticides can have clear negative effects. Pesticides work by their toxicity to living organisms. As a consequence, not only can they endanger cattle, fish and other creatures but frequently also ourselves. Even used with care, it is impossible to prevent toxic chemicals spreading over the environment and inflicting harm, for example by accumulating in the food-chains. And, pesticides can even have adverse effects; for example, a pest develops resistance but its natural enemies do not and are eliminated; in this way the pest has actually been helped to multiply much more quickly.

In our view, the objections to pesticides are so serious that everything should be done to minimize their use. A combination of non-chemical crop protection methods such as crop rotation and the cultivation of resistant varieties would make this possible and pesticides could be

12

used only in the last resort. Emphasis should be placed on care for crops, so that spraying is reduced to a minimum. This brings a considerable drop in costs which as a rule signifies an increase in profits (extra yield minus costs). This set of rules is called Integrated Pest Management (IPM). The IPM-approach to agriculture more closely approaches farming methods traditionally practiced in the Third World.

1.2 What can the reader expect from this book?

This book is aimed at development workers in rural areas in the hope that with the information given they will be able to support small-scale farmers in developing a sustainable agriculture which provides a decent living. Naturally, this book is not intended as a handbook providing instant answers to major problems. The situations in the Third World are far too diverse for that. We do however wish to provide a number of important principles, which can function as basic information for later on-site elaboration.

In the remaining chapters of part I the main emphasis will be on the social background of crop protection in the Third World.
Part II is the core of the book. It is dedicated to alternatives of chemical crop protection currently used: Integrated Pest Management, and offers a variety of practical suggestions for reducing the use of pesticides.
Numerous cultural practices for keeping pests at an acceptable level are discussed as are a number of suggestions for biological control. We also provide information about the safe and efficient use of pesticides: "if you have to do it, do it right". Some light is also thrown on the protection of harvested products.
Once it is known what can be practically done to reduce the use of pesticides, it is then a question of how to pass this knowledge on to the farmers. Part III deals with possible methods for training in IPM. Part III also provides some background information about research on IPM.
Many pesticides used in the Third World are forbidden in the West as being either highly toxic or persistent and therefore too dangerous. Moreover, pesticides are often sold without sufficient instructions for the user. For this reason Part IV looks more closely at government measures and legislation with regard to pesticides in the Third World. The Code of Conduct of the FAO (Food and Agricultural Organization) which lays down proposals for the use of pesticides is discussed, and could serve as a guide when judging existing regulations.
The text closes with part V containing two case studies about crop protection in Sri Lanka and Peru which enlarge on a number of aspects touched on in the first four parts.
Finally, four appendices are included. The first is a list of the 50 most common pesticides together with their corresponding dangers for health and the environment. Appendix II provides a list of organizations involved in the area of crop protection and Appendix III is a literature list complete with a short description of the contents of each book. In Appendix IV a list of words of phytopathological jargon is inserted.

2. Social background of crop protection

This chapter describes the transition in the Third World from traditional farming practices to an agriculture which makes use of modern technology, in particular pesticides. It discusses the compatibility of modern crop protection with traditional farming methods.

2.1 Background to the "need for development"

Developments in crop protection, and in agriculture in the Third World more generally are characterised by the import of known and successful models from the West. This can be explained by the Western faith in linear progress, which has so deeply coloured the Western world vision, that even the great opposition movements of the 18th and 19th centuries, the Enlightenment and Marxism, had no choice but to accept this dogma as entirely obvious. There would be an unequivocal development in world history: we are all heading to a kingdom of heaven, secular or otherwise, only the West is way ahead of the "others". If the "others" would only make an effort and follow the guiding light of the West, they would make it in the end. Of course, there are also Western traditions that consider the "others" as essentially different: fascist and related ideologies see them as necessarily inferior, while more romantic traditions consider them as superior because of their close relationship to nature. Anyway, it cannot be denied that faith in progress has often been and still is tainted by racist ideas. The place of a civilization on the evolutionary ladder of development in this world is largely determined by what that civilization achieves in the material sense. At the top of the ladder stands the Western World and right at the bottom the "primitives" such as the Aboriginal and African tribes.

Even now ideas about international cooperation are still largely determined by such evolutionary theories. In the past, voyages were made from Europe to the Third World to civilise the primitive natives and to convert the heathens; nowadays they go there to help the underdeveloped. In that respect, little has changed: Western civilization is still the standard by which other development levels are judged.

The message of this book does not come from the heart of this attitude but neither is it entirely free from it. Whether we want it or not, the influence of the West is manifest in many parts of the Third World, with positive but also with sorrowful results. This book has been written to reduce the excesses of this influence in the area of crop protection. It must give a realistic alternative which relates to small-scale farming **already influenced by technology from the West**. It is not intended to be a handbook for the environmental sound and human introduction of Western agricultural methods in those rare regions where agriculture is little influenced. Neither is it an introduction to a pesticide-free agriculture, the use of pesticides is too common to be denied, even though we see this as the final objective. But we see integrated pest management, which implies cautious use of pesticides, as being a good step in this direction.

Figure 2.1. Recent developments in "development work".

2.2 Small-scale farmers: an unambiguous concept?

As do many publications, this book frequently uses the term "small-scale farmer" to define its target group. However, it is not always clear which group is intended by this term. Does "small-scale" refer to the size of surface area being cultivated, or to the level of income, or does it refer to the degree of cash crop cultivation? Even the term "farmer" is not as unambiguous as it sounds. Can someone who only spends 50% of his time in agriculture be called a farmer? And how do you define a nomadic cattle keeper or the owner of a lemon orchard of only 2 hectares? Does the term "farmer" imply a man or a woman, bearing in mind that in Africa 75% of all agricultural work is carried out by women?

The most common agricultural system is the small scale family agriculture with a farm of relatively limited area which markets a part of its agricultural produce. Modern technology, in particular pesticides and artificial fertilizers, are used to a lesser or greater degree. A transition can be seen from traditional, subsistence farming towards market oriented agriculture. This combined use of traditional knowledge and methods with Western knowledge and techniques means that farmers are being confronted by the problems which the new technologies bring with them, especially in areas of public health and environment. Alongside this there are the problems typical to a change-over from one system to another. We try to analyze the transition from a traditional to a market oriented system. We begin by outlining the traditional farming methods and then move on to describe what measures are being introduced to achieve 'modernization'.

Four characteristics are used to describe transition from traditional farming to a market oriented agricultural system:
- marketing of agricultural products as a decisive income factor;
- production means as a decisive cost factor;
- crop protection technology, an important factor for the level of production;
- knowledge of crop protection, an important factor in determining to what extent the concept of modern crop protection is appropriate for the small-scale farmer.

In the next paragraph, traditional agriculture will be described. After a paragraph on modernization of traditional agriculture, some characteristics of 'modern' farming will be elaborated.

2.3 Traditional agriculture

Marketing of agricultural products. On the whole, most of the crops cultivated are for domestic use; any agricultural surplus is sold in the local or regional markets. In many countries, especially in Africa, the task of providing the family's food supply is left to the women, while the men are predominantly occupied with livestock production and cultivating cash crops. The small amount of income generated leaves little for purchasing seeds, fertilizers, pesticides, etc.

Production means. Very few of the inputs are imported from outside the region. Most of the varieties have been selected over a number of years from the best plants at harvest. Energy and other inputs, such as fertilizers and pesticides, are not purchased but, where possible, on-farm produced.

Crop protection technology. Of first and foremost importance is that traditional crop protection is not an independent series of treatments, but is related to other aspects of a culture. Traditional crop protection, therefore, cannot be viewed as separate from other cultivation

Figure 2.2.. Who is the small farmer?

Figure 2.3. A traditional landscape as drawn by a Peruvian farmer.

practices. Mixed cropping is one example of a traditional farming technique which is used to check the spreading of an infestation and to prevent possible crop damage. Mixed cropping is also important for increasing the plants' utilization of nutrients, for a more efficient use of the soil and for spreading risks. In addition traditional crop protection uses various preventive measures aimed at specific pests, like for instance removing or burning rice and corn stubble to control stem borers. Thirdly, traditional crop protection techniques originate in the regions where they are practiced. If pesticides are applied, these are generally produced locally. Often plants and plant extracts with specific special characteristics are used, for example repellents such as garlic and botanical pesticides (like Pyrethrum).

Knowledge of crop protection. In traditional farming systems, the farmers' knowledge of crop protection is usually amassed through the experience of many generations. This type of knowledge is slow to change and sometimes lacks insight into general processes. Of visible pests like insects and weeds it is often known at which stage it will cause damage, but invisible pests such as fungi, nematodes, bacteria and viruses are mostly unknown.

2.4 Transition from traditional to market oriented agriculture

Motivations and causes for change. What motivates a farmer who lives and works in a traditional situation to switch to market oriented, modern technological methods of agriculture? Partly, a farmer will start producing for the market from a desire to earn more money to buy consumer goods. However, it is not usually a matter of free choice but for example the result of a drop in regional prices because a larger farm has increased its production by using modern inputs. Also direct pressure plays a significant role. From long ago, farmers have never really

17

been free to utilize land as they saw fit. Rules have always existed for determining who could cultivate what, even for using which methods. Whether or not colonization by Europe is taken as a starting point, the strength of the central government is growing constantly, and all sorts of cultivation decisions are being taken out of the hands of the farmers. This widespread tendency is still increasing and is coupled with measures to harness part of the agricultural production for the government and the social groups associated with it.

Phenomena during the transition to modern technology. The institutions which used to regulate agriculture and the distribution of surpluses are being forced open. Poorer families in particular see their rights to use common grounds being endangered. Sometimes control over major pieces of land is removed from the village community or from small-scale farmers, if necessary with aid of military power, and given to selected families to start farming. The cultivation of particular crops is frequently imposed. The traditional division of labour between men and women is also under serious pressure. Where women used to be responsible for food-crops and men were occupied more with animal husbandry, in a number of countries this is now being turned upside down; more or less by decree. Men are appointed to be "the farmer" and are put in charge of the inputs supplied. The result is usually that these inputs are used solely for cash crops, by tradition a man's job, and that the profits are spent on luxury goods and social contacts. A decline in the family food supply is a common symptom of such interference.

Origin and objective of modern technologies. Modern technology is predominantly developed by manufacturers of inputs such as pesticides. These manufacturers have the necessary facilities and the personnel to undertake research, giving them major control of technological developments in agrarian production. They target themselves towards the West as that is the market with the most capital.

In Western countries expensive labour is replaced by capital. Technological developments in agriculture are also aimed at reducing labour, through mechanization of the production process and for example through the use of herbicides. However, the Third World there is no shortage of cheap labour. There, modern chemical control technology only reduces the demand for labour in agriculture which leads to less employment.

From dependence on the environment to dependence on inputs and market. Modern technology attempts to increase the farmers' control of the cultivation process. The other side of the coin, however, is an increasing dependence on inputs from the market. We examine this point further by looking at the production processes which play a part in agriculture. Farming in a particular region can be seen as the sum of a production chain:

- production of and control over inputs (labour, land, seeds, fertilizers, pesticides);
- cultivation of crops;
- storage, trading and processing of products.

In traditional farming, the farmers control the entire chain themselves. Integration in the market is leaving fewer and fewer parts of the production chain under the control of the farmers. In fact, the supply of inputs and the marketing of products by dealers or cooperatives implies that only the cultivation process and its inherent risks remain.

Once a region is involved in the development towards a more market oriented agricultural system, it is almost impossible for an individual farmer to remain detached. Other farmers, extension officers and dealers exercise social pressure for the continued use of new techniques. A return to the traditional situation is also no longer a real possibility. For example, if in a region

wood			cellulose
sugar cane			sweeteners
soya	biomass		alcohol
wheat			starch
sorghum			chemical substances
maize			proteins

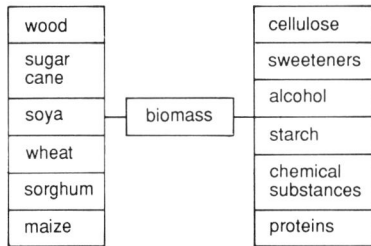

Figure 2.4. Fractioning and recombi-nationtechnology. Out of any agri-cultural product many products can be produced.

only improved, hybrid varieties are left, individual farmers are no more able to produce proper seeds. Furthermore, farmers may have debts to suppliers for inputs acquired from outside their farms, which can only be paid off by increasing production levels through the continued use of the new technologies.

Integration in the market creates competition with other regions, and consequently some crops can no longer be cultivated profitably. A region therefore has to specialize in a smaller range of crops which has its own risks. Involvement in the world market means competition with many other countries; the market of many unprocessed agricultural products is already saturated so that prices are often too low to even cover production costs. The risks of integration in the market are even greater for products as grain and sugar. Large scale dumping of such products on the world market by the USA and the EEC cause enormous disturbances of the market balance and cause price levels to drop way below production costs. Unfortunately, the Third World is forced to adopt an export economy to repay their enormous debts. It is often the small-scale farmers who are the victims of the crisis on the world market.

Developments in biotechnology play a role throughout all this. Important to producers of agri-cultural products is the technology known as "fractioning and recombination". This technol-ogy makes it possible to break down any agricultural product into a number of elements (fractioning). These elements can then be recombined at will to form semi-manufactured- or end-products (recombination). This process in fact reduces agricultural products to biomass (fig. 2.4). They lose their specific significance for the consumer market and the consumer pat-tern.

A well-known example of this is Coca Cola's switch from cane sugar from the Third World to isoglucose made from corn, which practically meant an economic death-blow to small-scale far-mers in the Philippines. The ability to make such switches gives an enormous amount of flexi-bility to the agri-business. Relative price ratios and certain technological conversion keys can be used to play off the production sectors and production areas one against the other.

2.5 Agriculture influenced by Western technology

Agriculture nowadays is still changing towards a market oriented business, following the above described processes. Of course, the greatest part of the Third World agriculture is somewhere be-tween traditional and fully market oriented agriculture. Some characteristics of a modern agri-culture are described below.

Marketing of agricultural products. Marketing makes certain demands of the quality of the produce, its transport and storage, especially for vegetables and fruit. A market oriented produc-tion system can thus necessitate the control of pests or diseases, and as a consequence an in-crease in the use of pesticides. For example, the consumer demands good looking products, for which better prices can be obtained. It is, therefore, sometimes beneficial to control pests which deface or cause external wrinkling but which otherwise have little or no influence on the productivity of a crop.

This can create greater gaps between the smaller and the larger farms, because larger farms have easier access to modern technology.

Production means. Pesticides are not used by a farmer as a single product but as part of a "package" consisting of improved varieties, pesticides and fertilizers. If part of the package is

Figure 2.5. A traditional farmer and his problems.

Figure 2.6. A modern farmer and his problems.

not available, problems arise, for instance, if improved varieties are cultivated when no pesticides are available risks for outbreak of a severely damaging pest increase.

For many crops, the costs of pesticides are low in comparison to fertilizers. If investment on fertilizers is threatened by pests, then from a financial viewpoint it soon becomes desirable to spray. As a consequence, spraying takes place before there is any clear damage, and even without any immediate threat to crops. In this way, chemical control is used as a preventative measure. When the costs of pesticides are high, or chemical control causes risks itself, it is easier to convince farmers of the benefits of using alternative methods.

Crop protection technology. In modern agriculture, crop protection usually means the chemical control of pests. The availability of pesticides makes the (preventive) traditional methods less interesting, for they can always be used in emergency situations. To use pesticides correctly one must know at which level of damage ("threshold") it is economically justifiable to apply pesticides. But, even when thresholds are adhered to, there are generally a number of hidden "costs":

- the costs of damage to the environment;
- the costs of dependence on suppliers and creditors implied by the introduction of pesticides;
- macro-economic costs: the necessary inputs must be imported which means spending foreign currency;
- the costs of the disappearance of traditional methods or resistant varieties which could be valuable for crop protection in the future.
- the costs of loss of employment.

Knowledge of crop protection. The efficient use of pesticides requires a certain way of thinking. A number of skills, like recognizing the connection between pest and loss, or handling spraying equipment, is required for applying pesticides properly. It is difficult to obtain unbiased technical information, usually provided by middlemen or extension services. Information on the use of pesticides usually is confined to the description of steps to be taken for application of a chemical without background information. Moreover this information relates to a very limited degree to the every day world of the farmer.

2.6 The reverse of modern technology

In the last twenty years the attitude of the West towards pesticides has drastically changed and they are now used with much greater caution. This is evident from the fact that less persistent pesticides are being used, but it has not as yet led to a reduction in their use. The quantity of pesticides applied in Western agriculture, expressed in kilograms of active ingredients per hectare, is still increasing annually.

In the Third World, the dangers related to the use of pesticides are not yet generally acknowledged, and environmental problems are not usually given priority. Consequently, pesticides which have been forbidden or hardly are used anymore in the West are being dumped on to the Third World markets. This can be seen in the growing supply of broad spectrum and persistent pesticides which in the West are either banned or subject to strict control regulation. The financial position of small-scale farmers is also an important factor. Apart from the fact that modern, less persistent pesticides are not always available in the Third World, they are also too expensive for many small-scale farmers. Trapped between the financial limitation of

20

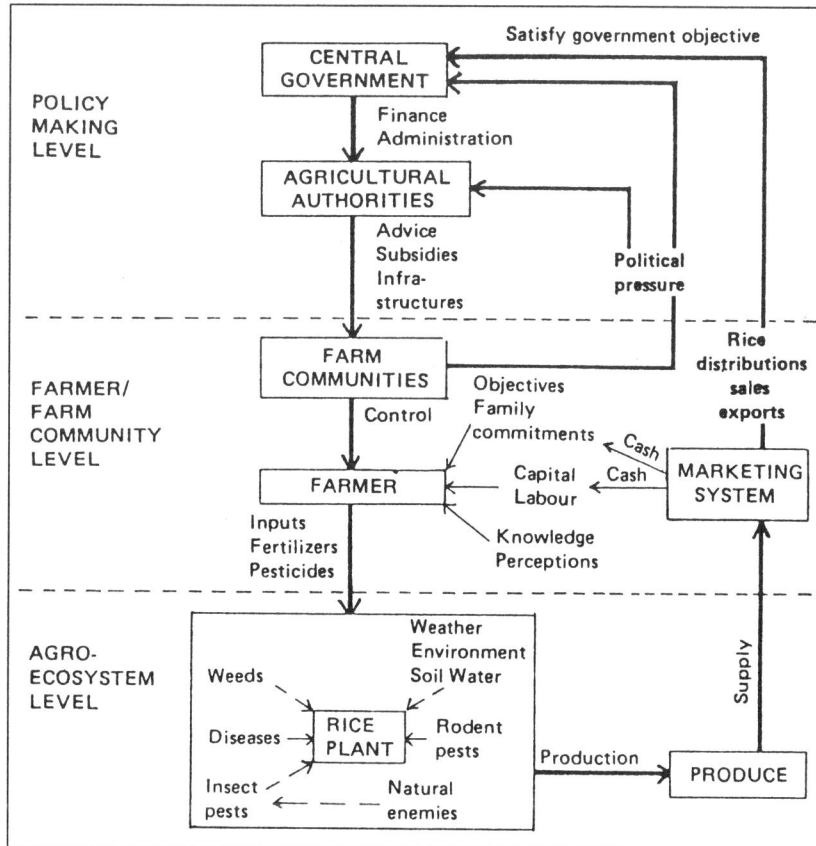

Figure 2.7. A rice production system in the tropics, and the place of pests in it.

the farmers and the information provided mainly by the agro-chemical industry, the extension services are yet unable to bring any real changes to this situation.

The latest developments, again propagandized in the West, strive to return to a form of agricultural in which farming systems are viewed as one integral operation. This research (i.e. farming systems research (FSR)) centres on the farm household and examine the influence of certain cultural practices on the total system.

For crop protection this means for example that prevention of pests is included as part of the initial planning. This research directive harmonizes with Integrated Pest Management in that it tries to minimalize the use of pesticides through the combination of methods. Thus, some attempt is being made to prevent the environmental problems now being experienced in the West from becoming manifest in the Third World.

The next part of this book looks at the subject of Integrated Pest Management in more depth. This method of crop protection certainly offers a number of significant advantages, however, in its application one should take into account the marginal position of the small-scale farmer in Third World. There must also be awareness for the unavoidable skepticism that this umpteenth new solution will invoke in farmers. After all they have, more than once, been offered the Ideal Solution.

Part II
Implementation of integrated pest management

Part II looks at the practical performance of integrated pest management. After a general introduction, it discusses cultivation practices, biological and chemical pest control within an integrated pest management system and closes with a chapter on storage methods. Part II thus has an agro-technical character.

3. Introduction to Integrated Pest Management

At first sight, a field appears to consist of a plot of land covered by a crop. Closer observation, however, reveals a more complex situation: every field contains an entire living community. Above and below ground it seethes with insects, mites, spiders, nematodes, bacteria and fungi, rats, weeds and so on. These organisms compete against each other for food or light, they devour each other, some species live together in order to survive. Most of the organisms have no appreciable influence on the crop being cultivated in the field. There is a number, however, which does influence crops; some favourably others negatively. The latter organisms deserve further consideration as far as pest control is concerned.

Figure 3.1. Many organisms live in a field with a crop, by far the most of them having no effect on the yield. Some of them however affect the crop, and these organisms, both beneficial and harmful ones, are of interest to the farmer.

3.1 Pests

Organisms which have a negative influence on the harvest are called pests. Pests can be insects, fungi, bacteria, viruses, nematodes, rats, birds or green plants. Damage which is caused by pests of microscopic size (fungi, bacteria, nematodes) or smaller (viruses) is also referred to as "disease", pests being green plants are called "weeds". Each of these attacks a field in its own manner. Insects suck or eat from the plants and bring over diseases. Rats and birds gnaw at or pick peck plants, sometimes clearing the way for other pests. Fungi and bacteria grow on

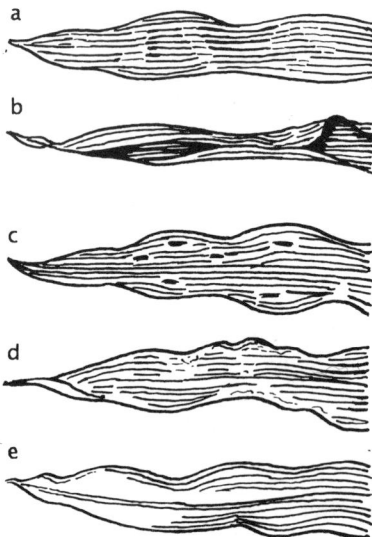

Figure 3.2. Some symptoms a plant can show. (a) a normal leaf; (b) a leaf which is dried out (greyish green colour); (c) a leaf infected by a fungus; (d) a leaf affected by a phytotoxic spraying (tips are burnt, tissues die, leaves become white-cap); (e) a leaf showing nitrogen deficiency symptoms (leaves become yellow starting at the tip).

Figure 3.3. The increase of a pest population during the course of a growing season. This increase can be expressed in several ways: by the number of individuals per unit of area (e.g. with insects, rats and nematodes); by the area of infected leaf area (e.g. with bacteria or fungi); or by the biomass (weeds). Any suggestion of a linear connection is incorrect; during the course of a season the increase of a population is generally exponential. The Y-axis should in fact be read as a logarithmic axis.

Damage and loss

Pests cause damage to the crop. However, damage does not necessarily lead to loss, a damaged plant may yield well in some cases. For example: a leaf-caterpillar in corn may gnaw away the tops of the leaves so that the crop looks terrible (see fig. 3.11 en 3.12.) Yet if the growing point of the plant is not injured, the actual harm done is negligible as the plant normally has more leaf surface than it can use. Therefore, even if a number of leaves are totally eaten away, a reduction in product yield is not the definite outcome.

Similar mechanisms can be seen at work within a crop; if one plant of a cereal crop is damaged or dies, its neighbouring plant is able partly to compensate by extra tillering.

or in the plant tissue, either consuming it, poisoning it or blocking its vessels. Weeds compete with a crop for water, nutrients and light.

Usually the population of a pest at the beginning of a growing season is small, but as the crop grows, it increases. This can be simply expressed in a graph (see figure 3.3).

The extent of a pest population when first observed and the rate at which it increases is decisive to assess the decrease in yield which could be caused. If an infection of aphids grows from 1 to 100 individuals per hectare, the harvest will not diminish significantly. However, if the aphid population increases from 1 to 100 individuals per plant, there will indeed be a serious harvest loss. For this reason, maximum tolerance levels have been established for a number of pests. When the population exceeds this threshold, the decrease in yield caused to a crop cannot be tolerated. The lowest population level at which unacceptable damage is caused is called the **damage threshold**. Figure 3.5 shows the graph from figure 3.3 with the damage threshold inserted. In the situation shown, the damage threshold is exceeded some time before the harvest and there will be a loss of yield.

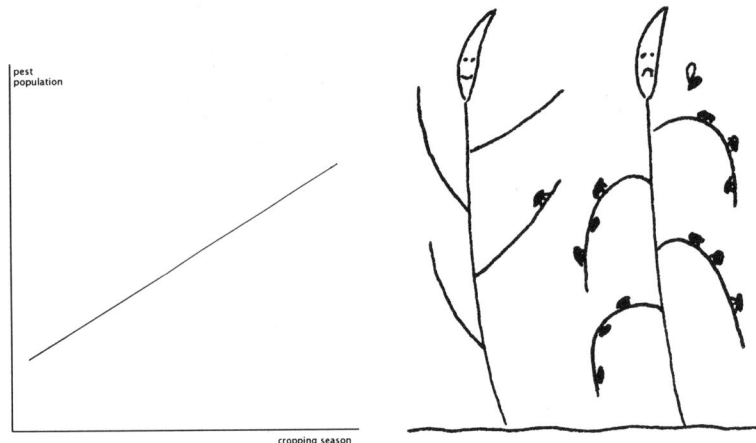

Figure 3.4. In most cases pest organisms are only scarce in a field, so that they do not do much injury. Sometimes, however, pests get so numerous that control may be necessary.

26

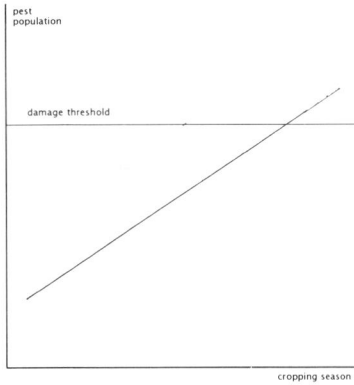

Figure 3.5. The growing of the pest population and the damage threshold. Here the pest population exceeds the damage threshold some time before the harvest. For a farmer this level of damage is unacceptable.

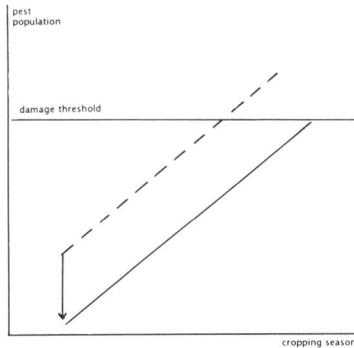

Figure 3.6. When applying measures of sanitation, the pest population is low at the beginning of the growing season. Consequently the pest stays at a low level throughout the season.

The level of a damage threshold depends on the crop, the pests and its position on the plant and on the time of the season the trouble-maker appears. A lesion on a leaf for instance, generally causes little damage; a lesion on the stem on the other hand can kill a whole plant. In practice there is a rule of thumb which indicates that a 10% reduction in harvest is no longer tolerable. If the damage is higher, measures against the relevant pests should be considered seriously.

3.2 Crop protection

The purpose of crop protection is to keep the pest below the damage threshold. There are three strategies to bring this about:
- ensuring that the pest does not get into the field (sanitation);
- ensuring that the pest has little chance to multiply;
- eliminating the pest.

A survey of these different strategies is given below.

Sanitation. By taking certain measures of hygiene it is possible to ensure that as few pests as possible get into a field. These measures should be undertaken especially at the beginning of the growing season, they can be ploughing under weeds and burning or adequately composting already infected plant remnants, taking care no infected planting material is used etc. When it starts at a low level it will take the pests longer to increase to the damage threshold level. The best possible result is either that it never shows its presence in the field at all (of importance for soil diseases), or that the pest never manages to reach the damage threshold as the crop has already been harvested before the population gets too big.

Sanitation is effective if the population at the **start** of the growing season is decisive for the damage caused, as with weeds and some insects. If crop usually is infected **during** the growing season, as with some insects and fungi, sanitation is not effective. Pests often reproduce rapidly in the tropics because of the high temperatures and therefore a small initial population can frequently be sufficient to cause the damage threshold to be exceeded during the course of the growing season.

Reducing the reproduction rate of pests. A second strategy is to reduce the rate at which the pest population increases. Many cultivation practices are applied with this intention (see chapter 4). For example: in many cases a crop which has been treated with a well-balanced manure has more resistance to pests than a crop which has been treated with an unbalanced manure. For certain reasons, pests multiply less efficiently in resistant varieties. Similar effects can also be expected from adjusted seeding- and planting-densities and mixed cropping.

Crop protection methods of this type, however, have often an ambiguous effect. A method which impedes the multiplication of one infestation could actually work to the advantage of another. For example, putting a lot of distance between the rice plants hampers the reproduction of the Brown Plant Hopper but increases weed problems.

Preventative chemical pest control can also be considered as an action for counteracting the multiplication of pests. Sometimes, the only remedy to losses is to spray a protective pesticide over the crop, in particular for fungal diseases in warm climates. For a period after spraying the crop is not susceptible to the fungus, which is not eliminated. Once the pesticide has disap-

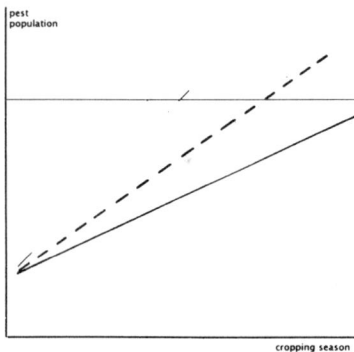

pest
population

cropping season

Figure 3.7. If the multiplication of a pest can be impeded, e.g. by applying cultivation practices which create an unfavourable environment for the pest, it is possible to keep it below the damage threshold.

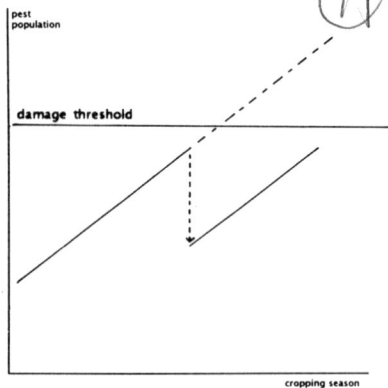

pest
population

damage threshold

cropping season

Figure 3.8. Destroying a part of a pest population threatening to exceed the damage threshold can prevent damage being done to the crop.

peared, or new leaves appear on the plants, growth of the fungus simply resumes (see figure 3.7).

Methods for killing a part of the pests population. A pest population threatening to exceed the damage threshold can be partially destroyed. This can be done mechanically, e.g. by hoeing weeds, or in the popular chemical manner. Curative chemical control usually is carried out against insect pests, weeds, and sometimes against fungi. If after spraying the pest population manages to multiply rapidly once more, one treatment is not enough and the will increase further (see figure 3.8).

Natural enemies frequently keep the pests at a sufficiently low level (biological control). A natural enemy can be in a field by nature (most damaging insects are assailed by all kinds of natural enemies), or they can be introduced by man. Biological control is mainly effective against insects, but sometimes also against nematodes and weeds. In the soil most pests are controlled in this way as well.

3.3 Integrated Pest Management

Integrated Pest Management (IPM) officially signifies: "The application of the best mix of environmentally sound techniques in order to keep pests below the damage threshold." In fact integrated pest management means avoiding dependence on one single method for controlling pests. Various measures are undertaken, reinforcing one another, to ensure that the pest does indeed stay below the damage threshold.

The ultimate objective is to reduce dependence on chemical control as the main input for crop protection (so to reduce dependence on the chemical industry and on hard currency). A cutback in the use of pesticides lessens environmental pollution and reduces the risks of poisoning the user and the consumer alike. Furthermore, applying IPM generally reduces the farmers' production costs since fewer pesticides are used, while it retains or increases the yield so that the position of the individual farmer improves.

Some important elements of IPM are: sanitation, the use of resistant varieties, preserving natural enemies of pests and the natural balance, and avoiding excessive manuring of the crop. Pesticides only being applied in case the damage threshold is exceeded.

28

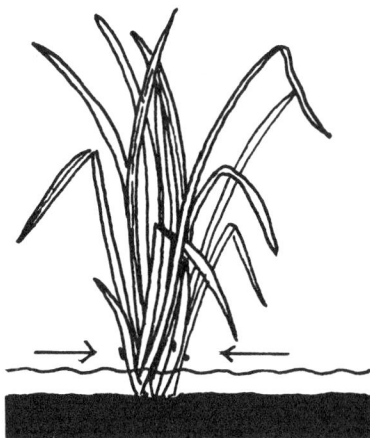

Figure 3.9. The brown planthopper lives on the lower part of the plant, so when spraying, the insecticide should be applied at the base of the plant.

Figure 3.10. The brown planthopper, the insect is about 3 mm long.

Example

*The brown planthopper (**Nilaparvata lugens**) is an insect pest found in rice. It reproduces by two or three generations per crop. If the population reaches high levels, a combination of overcast weather conditions and heavy nitrogen manuring causes a reaction called "hopperburn". The plants turn yellow and dry out which can lead to the total loss of a harvest.*

The IPM treatment against the brown planthopper in rice is carried out as follows:

- *planting the rice over a whole area simultaneously to ensure that there are no fields where the population can start to build-up prematurely;*
- *burning off the stubble to destroy the brown planthoppers' overwintering sites;*
- *planting widely (15-20 cm) to reduce the reproduction rate of the brown planthopper;*
- *a well planned scheme for water-supply can destroy part of the brown planthopper population;*
- *the use of a manure low in nitrogen to prevent the crop becoming too dense thereby reducing the risk of hopperburn;*
- *cultivating resistant varieties to reduce the rate at which the brown planthopper multiplies;*
- *if the number of predatory spiders (natural enemies) is high in comparison to the number of brown planthoppers the crop should not be sprayed;*
- *if, however, there is a high number of brown planthoppers in comparison to the number of predatory spiders, spraying should be carried out and directed at the base of the stem where the brown plant-hopper is found, and not over the canopy, where natural enemies are found.*

By combining a some of the above mentioned measures, the number of sprayings can be halved, while the harvest will remain the same or even increase.

For efficient control of a pest it is always necessary to know with certainty which organism is causing the damage. This is not always simple.

Observed damage does not automatically mean, for example, that the beetle found on the leaf is the culpable organism. This beetle is perhaps able to survive perfectly well on caterpillars which come up from the bottom of the plant at night to chew at the leaves. Spraying in the daytime will in this example mainly injure the beetle and leave the caterpillar to doing its damage, and can have actually more negative than positive effects.

Deficiency symptoms or damage from cold are often confused with virus or fungal diseases. It is self-evident that the spraying of a fungicide will have no effect on a virus infection.

The rest of this part has been compiled as follows: each chapter deals with various methods and techniques each of which can contribute towards keeping pests below the damage threshold. In chapter 4 sanitation and cultural practices are dealt with. By incorporating cultivation techniques, an unfavourable environment for pests can be created. Sanitation keeps a crop as free as possible of pests. Chapter 5 deals with biological control methods: how can a natural enemy be introduced or preserved? In Chapter 6 chemical control is discussed. When is the use of chemical control sensible? What are the side-effects in and outside a crop? If applied, how can this be done as efficiently as possible?

Chapter 7 looks at problems concerning the storage of harvested products. Finally, Chapter 8 gives a survey of all measures which can be taken to deal with the various groups of pests.

Figure 3.11. Injury done at some parts of the plant may look terrible, but does not necessarily lead to losses. In case a caterpillar gnaws the upper leaves of a maize plant, the lower parts will catch more light and thus will compensate for the reduced photosynthesis.

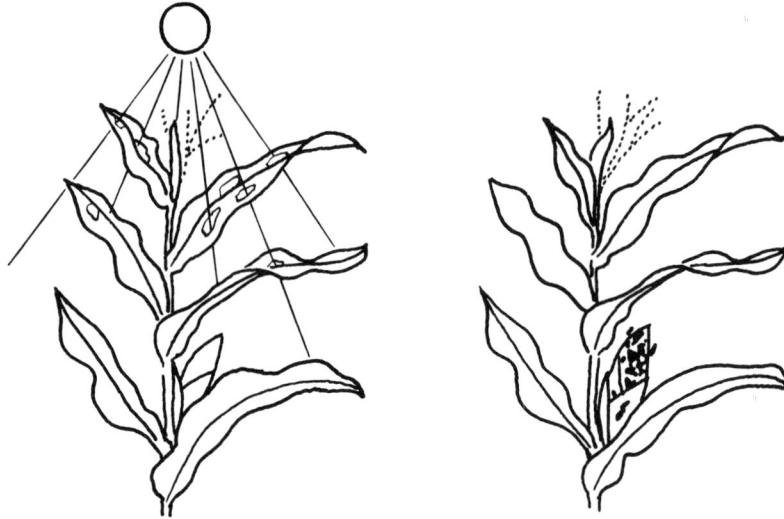

Figure 3.12. Injury done at the yielding parts of the plant easily leads to losses. Damage done to a cob therefore is harmful, because the yield is directly reduced in quantity or quality.

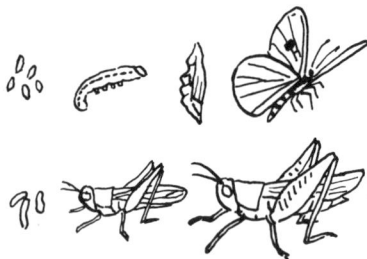

Figure 3.13. Most insects get through metamorphose during their lifetime. A butterfly and a locust.

Insects and mites

Contrary to what most people think, insects and mites do not belong to the same taxonomic group. *Mites* are related to spiders and scorpions and can be recognized by their eight legs. Spider mites (tetranychid mites) are important pests in several crops like cotton, tomato and fruit trees. They cause damage by sucking up the contents of leaf cells through their suctorial mouths, resulting in little yellow spots on the leaves. With high infestations the leaves get a brownish appearance. By then the photosynthetic capacity of the crop will be reduced considerably. An infestation of spider mites can also be recognized by web on the leaf surface. The mites itself are relatively small: about 1 mm. *Insects* have only six legs. Most adult insects have two pairs of wings. The adult lay eggs out of which larvae hatch. After a few larval stages (generally 4 to 6) again an adult comes into being. Some insect groups, such as moths, beetles, wasps and flies, have an immobile, pupal stage between larvae and adults. The larvae of these groups have a complete different appearance from the adults. Larvae from insect groups such as locusts, aphids, bugs are much alike the adults; only smaller and without wings.

It is estimated that approximately 6 million insect species exist, of which about 300 are pests for cultivated crops. Insects can cause very different types of damage. Many insect cause damage by eating leaf tissue (beetles, caterpillars, locusts). Others suck fluid from the plant causing growth distortions and often these sucking insects transmit viruses (aphids, plant hoppers, bugs). Furthermore, some insects cause damage which looks like a disease, for instance stem borers. The larvae of the latter bore themselves in the stem, so that the panicle dies ("white heads" and "dead hearts" in rice). Also insects, mostly beetle species like the corn-weevil, can cause great damage to stored products.

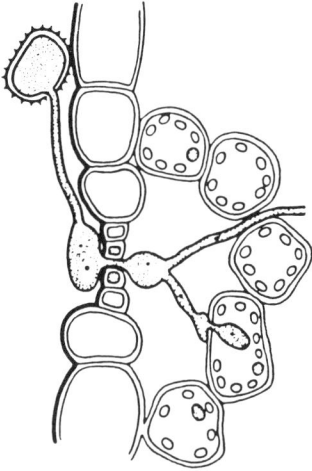

Figure 3.14. A fungal spore germinates and penetrates a stoma of a leaf. Examples of fungal diseases are rusts and mildew on various crops. Magnif.: ± 1000x.

Nematodes

Nematodes are little (ca. 1 mm), not-segmented worms, which usually are present in enormous numbers in the soil. Some species are damaging because they transmit viruses. Most nematodes cause damage by sucking on the roots, some attack the stems or cause wilting (for instance "red ring disease" of coconut).

Economically harmful nematodes are cyst nematodes in potatoes and root-knob nematodes.

Pathogens: fungi, bacteria and viruses

Fungi are plants without chlorophyll. This means they have to feed themselves with other organisms, living or dead. A lot of different groups of fungus species exists. The most familiar group being those with mushrooms. Mushrooms can be compared with the flowers of green plants; they produce spores out of which new fungal hypha (threads, the form of most fungi) grows, which again can produce mushrooms. Most plant pathogenic fungi produce their spores in another way, just think about white powdery mildew or brownish rust on the surface of infected leaves. About 200 important crop damaging fungal species exist. They can cause very different types of damage: change of colour (yellowing, browning), scab, growth distortions, cancers, wilting and rot. Some are relatively easy to recognize as fungal infestations. For example, when a part of the fungus lives on the leave like mildew or rust diseases. Other diseases are very difficult to diagnose. Is the bad state of a plant caused by a wilting disease, shortage of nitrogen or water, a nematode infestation, or a virus disease? Sometimes this even can be a problem for a crop protection expert.

Bacteria are small organisms with only one cell and can only be seen through a microscope. Few bacteria are plant pathogenic. It is not true, as is generally thought, that bacteria can only cause rot. Bacterial diseases can cause the same damage as fungal diseases. Only there are much less bacteria than fungi causing diseases.

Viruses are even smaller than bacteria (0.015 to 0.3 micron); they can only be seen with an electron microscope. Viruses are not organisms, because they lack any kind of metabolism. They consist of a nucleic acid molecule which is "wrapped" in protein. For their propagation they are dependent on living cells. Outside cells viruses are completely inert.

Viruses can infect new plants through seeds or seed tubers, directly through contact between plants, or indirectly through vectors. The most important vectors are aphids and plant hoppers, but viruses can also be transmitted by nematodes, spores of fungi, mites, mothflies, and beetles.

Symptoms of virus diseases are: discolouring (often mosaic), wilting, necrosis (dying of parts of the tissue) and growth distortions. Virus diseases cannot be controlled by pesticides. Viruses are present in plant cells and the functional elements of viruses are much alike those of cells (i.e. nucleic acids) thus the chemical that would "kill" the virus would inevitably also destroy cells of the organism itself. Control of viruses must be directed at the vectors or at prevention by using virus-free seeds and seed tubers. Also important is the search for resistant varieties, but that is equally true for all pests and diseases.

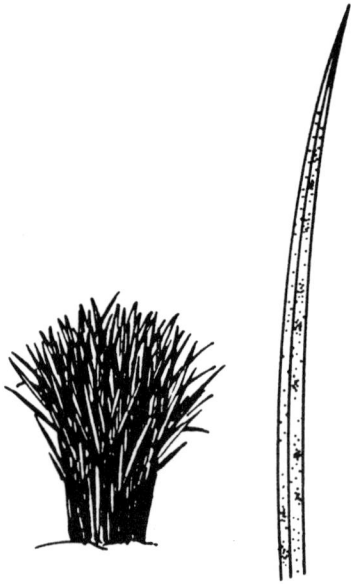

Figure 3.15. Plants infected with a virus usually get a strange appearance, while their leaves show discolourations. This plant shows symptoms of the grassy stunt disease: the plant does not grow well and tillers excessively, while leaves are narrow and stiff and show spots.

Figure 3.16. Weeds are green plants competing with the crop for water, light and nutrients.

Weeds

Weeds are the most harmful pests and this makes weed control an important factor in determining the cropping system. For instance, the cropping pattern is often largely determined by the prevailing weeds in a certain area. Also transplanting (in stead of direct sowing) is a weed control measure and of course ploughing and weeding are the most labour intensive practices in a farming system.

Weeds are plants which in a certain situation interfere with a person's plans. An example to elucidate this definition: when potatoes are grown, we think it to be useful plants. The next year between cereals, however, we consider potatoes as weeds. Weeds are damaging because they compete with the crop for light, water and minerals and this reduces the yield considerably.

Weeds can also cause damage because their presence lowers the quality of a product, makes the harvesting more difficult, blocks waterways (water hyacinth) or some are even parasites (witch weed *Striga* on cereals).

Weeds can be classified on several ways into groups. Two practical methods of classifications for the control of weeds are:
- shape of leave:
  - narrow-leaved *(monocotyledonae)* of which young seedlings have only one leave (grasses, sedges, bamboo);
  - broad-leaved *(dicotyledonae)*: seedlings have two leaves.
- annual or perennial: annual plants propagate only through seed, while perennial plants can propagate through seed as well as through tubers or rhizomes. Especially weeds with rhizomes can be very difficult to control (e.g. alang-alang).

32

4. Cultivation practices

Farmers consider many factors when deciding how a crop is to be cultivated, and they often carry out experiments in this field theirselves. Most cultivation practices have a wide range of effects. This chapter looks at the various cultivation practices as they generally affect the health of a crop. Phytopathologically cultivation practices usually are ambiguous; a change in cultivation may control one pest and at the same time encourage another.

Introducing a new cultivating practice usually is a complicated affair. The proposed changes are often not only of consequence to pest control, but also, for instance, affect the division of labour; water requirments of a crop; the quantity of seed needed, etc. A valid question is: why, in view of the potential increase in yield, did the farmer not implement the changes earlier? What circumstances are changed, so that an adaped practice will be beneficial?

This chapter begins by discussing various general practices and continues by following the course of a growing season in order to indicate where and how certain specific cultivation practices may influence pests.

4.1 General practices

4.1.1 Soil tillage
Tillage is carried out for obtaining good seed- or plant-beds. The phytopathological reason for ploughing is controlling weeds, and pests which remain in plant remnants. When turning the soil, insects are surfaced which pupate in the soil; whereupon they either dry out or are eaten by birds.

Example

*The caterpillars of the armyworm **Spodoptera** feed on the leaves of corn, rice, sorghum, cotton and other crops. As these caterpillars pupate in the ground, ploughing damages the pupae or causes them to surface where they dry out or are eaten by birds.*

Example

*Host plants of "Chafer Grubs" **Schizonycha** are corn, wheat, sugarbeet, sorghum and groundnut. The larvae live in the ground where they inflict serious damage on the roots of plants. Deep ploughing destroys many pupae and larvae, either killing them directly or indirectly by exposing them to sunlight and to predators such as birds.*

Figure 4.1. When ploughing, pupae may surface and dry out. Other soil-borne pests are also influenced by tillage.

33

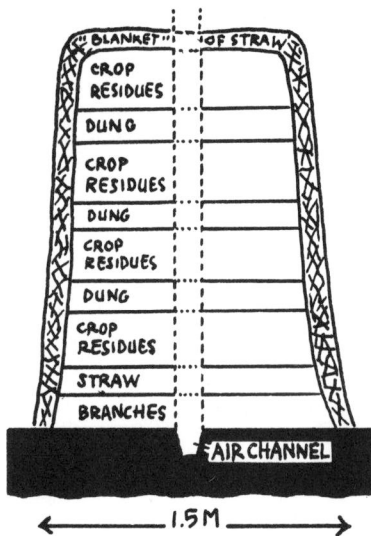

Figure 4.2. If infected crop residues are composted well, pests are eliminated and compost can be used safely as fertilizer. This is an example of a compost heap, with an air channel under and through the heap (a bundle of branches or wire netting). The heap is covered with straw so that the outer layer is also warmed up to 50 °C or more, so that most pests are killed.

Silty and clayey soils suffer from intensive tillage due to loss of structure (e.g. ploughshoe and crust formation) resulting in a reduced infiltration of water. On slopes tillage involves risks for erosion. For these reasons, and because of the high costs of ploughing and harrowing, there is an increasing tendency to employ "minimum tillage" in mechanised agriculture. This method involves placing manure and seed into small ditches. Herbicides are then used to eradicate the weeds.

In non-mechanised agriculture tillage is generally minimal. This is labour saving and anyway the machinery required for intensive tillage are often not available. However, in order to control weeds superficial tillage of the whole field is still required, otherwise intensive weeding must be carried out. Some weeds, specially grasses which form underground stems, are almost impossible to eradicate unless the undersoil is completely turned.

Countering the advantages of mechanical tillage, ploughs and tractor wheels carry soil from one place to another and thereby facilitate the spreading of a pest over the whole field. Therefore machinery and tools should be cleaned when going from one field to the other.

4.1.2 Hygienic measures

The most important way to prevent losses from pests is carrying out hygienic measures.

At national or even continental level carantaine measures are often taken. That means that imported plant material is disinfected, or stored until it is clear that it is not infected with a pest. Numerous pests have been and still are being imported from one continent to the other, usually causing huge losses. For the enthusiast traveller this means: never take plant material from one country to the other, unless assisted by an expert. By the way, in most countries it is strictly forbidden to import plant material without control.

At local level the goal of hygiene is to get no extra pests in the field, or to reduce the pests present before the crop is being planted or seeded. Some simple measures should be carried out in order to clean the field or prevent it from being infected:

- Always try to be sure plant material or seeds are free from pests. Especially the roots of seedlings may carry soil diseases as nematodes or soil fungi. In case of uncertainty, chemical disinfection of the seedlings may prevent the field from being infected.
- When applying manure, try to be sure it is not infected with weed seeds. Some seeds are able to survive a trip through a cow's belly. Proper composting of manure will kill them. Compost of infected plant material only is safe when the composting temperature exceeded 55 °C for some time. Usually, when the outer 10 cm of a compost heap is removed or a blanket of straw is used to isolate the heap, this temperature is reached when composting properly.
- Be careful when going from one field to the other not to carry soil. Therefore it is advisable to wash boots, tools and machinery regularly, preferably every time when changing fields.
- Stubble from previous crops should be burnt, grazed or ploughed under. In cereals in particular, stem borers tend to overwinter in stubble, and aphids overwinter in young sprouts which may appear after the crop has been harvested.

Figure 4.3. The farmer should take care not to spread pests from one field to the other. Pests are spread by dirty tools or boots, or by transplanting seedlings from infected fields.

Example

Well manured young coffee plants are more resistant to the nematode Meloidogyne spp than unmanured plants. The proper amount of manure allows the roots to develop a physical barrier against the nematodes.

Example

Root knot nematodes Meliodogyne spp cause severe damage to vegetables. They can be controlled by spreading animal manure on the seed beds. This manure contains fungi which attack the nematodes.

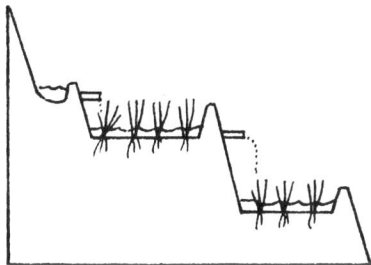

Figure 4.4. Basin-irrigation enables the farmer to execute a special irrigation scheme, by which pests like weeds or stem borers may be controlled.

Figure 4.5. When applying gravity-irrigation, incidence of pests may be influenced by varying the duration of the dry intervals between water supplies.

4.1.3 Manuring

Well-balanced manuring results in healthy plants. Usually good manuring raises plant's resistance against pests, although some pests and weeds are favoured then. Please note: well-balanced manuring is not the same as excessive manuring. For many crops excessive nitrogen especially increases susceptibility to pests. Organic manure has the advantage over artificial fertilizer that it strongly stimulates soil life. This has an impeding effect on some pests.

4.1.4 Irrigation

Irrigation has radical consequences not only for the crop but also on the entire farm system. For irrigation, the farmer's presence in the field is needed at certain times and the fields must be either levelled or with uniform slopes. An irrigated cropping system usually shows different mayor pests than a rainfed system. There are three important forms of irrigation:
Basin irrigation, used to irrigate levelled fields, the whole field is set rapidly under water, shutting the supply off and allowing the water to infiltrate into the soil. This form of irrigation is used frequently on heavy soils in delta regions, in orchards, where a small basin is made for each tree, and in wet rice cultivation.
Basin irrigation offers good opportunities for manipulating pests. If the ground is levelled sufficiently, it is possible to obtain a layer of water of equal depth over the whole field. Well thought out and planned, this method of flooding and drying out the soil can be used to control many weeds and pests. The most appropriate planning for obtaining the best results depends on local conditions and the mayor pests. Using the basin irrigation method many weeds and insects can be drowned by irrigating the field before sowing, or be dried out by keeping the soil dry for some days.

Example

The caterpillars of the "Rice caseworm" use the rice leaves for making cocoons. It is a matter of life or death for the pupae that the coccoons are filled with water. Consequently, if the fields are allowed to dry out for two or three days, the pupae will die.

If the water depth using the basin irrigation method is allowed to reach 15 cm, most weeds (particularly grasses) will be controlled. Alongside this treatment, the floating fern "Azolla" can be introduced as a green manure between rice, as often happens in South East Asia. When the surfaces on the water is closed by Azolla, the fern absorbs nearly all the light, further diminishing the chances of weeds.
Gravity irrigation is employed in fields with uniform gradients, where water is allowed to stream from an irrigation channel into a field for as long as is needed to wet the root area. The excess water collects in ditches at the end of the field. The water can be led through furrows (furrow-irrigation), then the crops grow on ridges; or it can be allowed to flow freely over the ground (border irrigation).
Gravity irrigation offers fewer possibilities for crop protection than basin irrigation. The soils where gravity irrigation is applied is generally light, and consequently the water is quickly absorbed. The water depth is rarely more than 10 cm. As with furrow irrigation, the ridges are never submerged and weeds and other pests are therefore able to survive.

Tomatoes, tobbacco, cabbage, pepsi-cums, beans, etc. are averse to wet collars. Should the collar get wet for a long period, fungi (i.e. Rhizocto-nia, Fusarium etc.) will cause a rapid "damping-off" of the young plants and the slow death of older plants. So when transplanting the plants, care should be taken not to place them too deep in the ridge.

Figure 4.6. By sprinkler irrigation some pests may be washed from the leaves, on the other hand mud may splash on the plants by which the plant can get infected.

Example

*One way in which the bacterial disease **Zanthomonas campestris** in beans is spread is through the soil. Rain or irrigation water dripping onto the ground and causing mud splashes on the plants can encourage the spread of the bacteria. This is more likely to occur where there is low density coverage and a lot of bare soil between the plants.*

With irrigation by gravity the interval of floating the field is the most important variable factor in the control of pests. Depending on the situation, a longer or shorter interval can have a negative or positive effect. Longer intervals ensure that the ground dries out thoroughly, thereby causing the death of many pests. Water stress however, also affects the crop and this can make plants susceptible to a number of diseases. When executing furrow irrigation it is important for many crops that the collar (the place where stem and root meet) remains dry. This prevents many diseases.

Sprinkler irrigation, "artificial raining", may spatter the plant with earth, in particular when the canopy is not yet closed. This can encourage infections of the leaf and the collar. If pesticides are used, their application should be adjusted to the time of irrigation to avoid the pesticide being rinsed from the leaves too quickly. Sprinkler irrigation does, in fact, impede some pests by washing them from the leaves. For example, thrips can be kept to an acceptably low level by applying sprinkler irrigation.

Irrigation as a supplement to natural rainfall lessens the hazards of irregular rainfall and gives greater freedom when selecting a date for sowing. When irrigation is used to allow a certain crop to remain in a field for a longer period, or even for a whole year, the population build-up of a pest is not interrupted by a dry season. Intensive pest control then may become unavoidable. During the dry season, irrigated fields can also serve as a temporary residence for pests which normally infest wet season crops. This can hardly be observed on the irrigated crop. However, as the natural reduction caused by the dry season has been negated, the normal crop suffers much more from the pest.

4.1.5 Shading

Shade trees can be planted on purpose, or deliberately left over when clearing a forest. Shadow is also often created above seeding beds to protect young plants against sun scalding by stretching out canvas or placing canes or twigs. Particularly where manuring is not at an optimum, plants such as tea, coffee and cocoa can profit from the protection against sunlight. A disadvantage is that shadow trees use valuable space in a field and compete with the crop for water and nutrients.

Shade trees provide nesting places or shelter for natural enemies of pests, in particular for birds and spiders. Shading, however, does not have a clearly positive effect on crop protection; rather the effects are negative. It influences the microclimate within the crop. In the daytime the moisture content in the atmosphere inside the crop is higher and the temperature lower than in the full sun; at night the temperature in the crop is rather higher than outside the crop. Via the changed microclimate it can encourage fungal diseases and some pests and so necessitate the use of pesticides.

Example

*The natural enemies of the beetle **Hypothenemus hampei** do not feel at home in coffee shrubs because the light is inadequate. If natural enemies are to be able to control this beetle sufficiently it is important that shading be used with care, for example by pruning the plants and by pruning or removing the shade trees.*

Figure 4.7. A crop rotation scheme, as found on unirrigated fields in Nepali hills.

(a) In wintertime mustard is grown.

(b) Then maize is intercropped with french beans. Both crops cover the soil, so weeds are suppressed.

(c) When the beans are ripe, they are harvested and millet is planted between the maize, so that weeds get no chance. During the growing season of the millet the maize is harvested.

(d) In wintertime the land is left fallow, cattle grazes weeds from the land and soil fertility gets the chance to recover slightly.

(e) Next year maize is intercropped with potatoes. The soil is quite clean from potato-diseases because last few years no potatoes were cultivated on the plot, and weeds are suppressed.

4.2 Measurements before the growing season

4.2.1 Crop rotation

Most pests can only survive on a limited number of, usually related, plant species. Weeds also, are sensitive to the type of crop in which they grow: each specific crop favours the growth of certain weeds and prevents the growth of others. Crop rotation makes survival for these pests more difficult. For example, cereals can be alternated with beans or peanuts, sweet potatoes with cotton and cassava. Important factors in crop rotation are variety of the crops in soil coverage and in rooting depth.

Cultivating the same crop for two years or longer is likely to encourage a whole series of pests, usually of a sort difficult to control. A period of lying fallow interrupts the build-up of soil-borne diseases and gives the natural fertility of a field more chance to recover after a period of intensive cultivation. From earliest times, crop rotation has been the most important technique for keeping the soil healthy, and it is frequently the only one. For this reason, even in technically sophisticated agricultural systems where artificial fertilizers and pesticides are freely available, crop rotation remains very important. The one major exception to this rule is irrigated rice.

There are still further considerations for instituting a system of crop rotation. Apart from controlling soil-borne diseases, it is essential that the structure and fertility of the soil is maintained.

Beneficial effects of crop rotation are partly reduced if the susceptible crop comes back in the cropping pattern, either as a "weed" or in mixed cropping. It is thus important that seeds and tubers are carefully harvested so that in the following season they do not turn into plants on which pests can survive.

If a new cultivation practice, artificial fertilizer, or a new crop is introduced in a region, it sometimes happens quite unexpectedly that crop rotation is no longer adequate as pest control. The new pests must be identified and a suitable method of control developed. Consideration should be given to such remedies as adapting tillage, sanitation or chemical control, or to the introduction of a more resistant variety. If these measures prove to be insufficient, little else can be done than to restrict the cultivation of the susceptible crop. This can be economically very painful.

The Incas in Peru passed a law that potatoes must not be grown on the same land more than once in 7 years. Cyst nematodes proved to diminish in this period to such a level that the following potato crop produced a good yield again. Repeal of this law by the Spanish colonial power had disastrous results.

Planning the most suitable crop rotation scheme in order to avoid economical losses is a long and cumbersome process. It means testing cropping patterns for years, and can only be done, in fact, in well-organized testing stations which are able to continue tests for a long time. From a desire to increase the surface area dedicated to the cultivation of certain cashcrops, governments sometimes have forced farmers to put aside the local crop rotation schemes. The results in terms of pest control have often been disastrous.

4.2.2 Mixed cropping

Mixed cropping implies simultaneous cultivation of two or more crops in one field. In some cases the different crops are planted at roughly the same time, but it is also possible to plant the second crop during ripening of the first one. The latter method is known as "relay cropping". Since olden times the mixed cropping system, particularly for foodcrops, has been applied frequently by small farmers.

There are a number of advantages attached to mixed cropping. Firstly, because each type of plant grows to a different level both above and below ground: soil coverage is better and greater volumes of the soil are rooted. Secondly, mixed cropping spreads risks for a bad harvest: if for some reason one crop does less well, the other is likely to benefit from the extra space and water. The total yield per hectare is usually greater from mixed cropping than from crops cultivated separately.

The improved soil coverage is important for weed control. Corn, sorghum, non-irrigated rice and comparative crops are bad soil coverers and weeds are able to get a strong hold between the individual plants. Mixing these crops with sweet potatoes, beans or alternative plants with good soil coverage will largely solve this problem. Moreover, the protection of the ground against direct sunlight and rain increases, so that erosion is impeded.

The microclimate prevailing in a mixed crop is different from that in separately cultivated crops. The effects of this on a pest can be positive or negative, depending not only on the pest, but also on the reactions of possible natural enemies.

Example

*Farmers in the Philippines were able to control the corn stalk borer **Ostrinia furnacalis** by mixing the corn crop with ground nuts. This increased the number of predator spiders of the Lycosa species, which attack stalk borers.*

Example

In Haiti disease free cultivation of the pigeon pea has only been made possible by interplanting tall sorghum. This inhibits insects transferring the viruses, thus protecting the crop from virus diseases.

Figure 4.8. Mixed cropping influences development of pests in several ways. By better coverage of the soil weeds are suppressed, natural enemies of insect pests get better chances to hide, while spreading of all kinds of pests is inhibited.

Crops which botanically are not related are good combinations for crop protection; pests harmful to the one crop will usually leave the other alone. Of some crop combinations of related crops, however, a number of harmful effects are known. For example corn or sorghum mixed with barley is conducive to the barley yellow dwarf virus. The influence which various crop combinations can have on pests is complex, and every situation should be tried out in testing stations preferably in cooperation with enthusiastic farmers.

4.2.3 Selecting a variety

Cultivating resistant varieties is the most direct and safe method for preventing damage by pests. However, the free choice from the various varieties differ greatly from crop to crop and from region to region. Sometimes farmers use seed they harvested from the last season's best plants, but if seed has to be bought, the choice is often limited to what is available in the particular region.

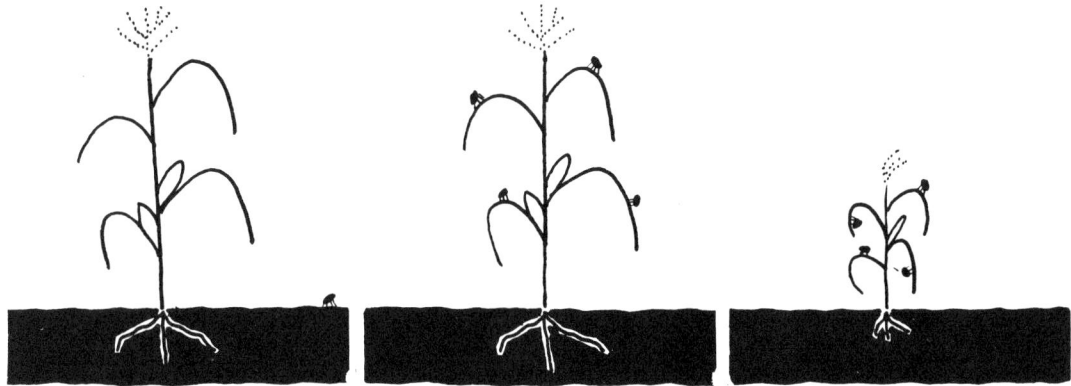

Figure 4.9a. A resistant crop does not allow a pest to live or multiply on the plant, so it can do no damage.

Figure 4.9b. a tolerant crop allows pests to live and multiply on the plants, but it is not affected by the pest.

Figure 4.9c. A susceptible crop allows pests to live on the plants, and in case of severe infestations losses will occur.

To make a valid comparison between the varying degrees of susceptibility of the different varieties to a particular pest, it is important to distinguish between the terms "resistance" and "tolerance".

In the case of **resistance**, a pest cannot live on a plant. For example, plant hair hinders an insect to walk or attempting to feed on sap, or by secreting poisonous chemicals a plant can defend itself chemically against a pest. A variety resistant against important soil disease(s) will not allow the build-up of a pest population and it is consequently very useful if the same crop is again cultivated on the field.

Tolerance on the other hand relates to the damage caused by a pest. The organism lives and multiplies on the plant but the actual effects on the harvestable part (fruit, grains, pods, etc.) is limited. A field with a tolerant crop can, however, become a source of infection for other fields containing more susceptible varieties, for the disease is able to multiply on the tolerant crop. In this case no susceptible follow-up crops can be grown.

A distinction can be made between so-called *horizontal* resistance and *vertical* resistance.

Horizontal resistance means that the plant is moderately resistant against a pest. A new line of the pest will not be able to break this resistance. This type of resistance depends on a large number of genetically determined characteristics.

Vertical resistance results in 100% resistance against a pest. This resistance usually depends on only a single genetically determined characteristic. The chance is great that a new line of the pest will be able to overcome the resistance and multiply, and the variety will be susceptible again.

Example

In Costa Rica cassava is planted in-between a corn crop which is almost ready for harvesting. The corn stalks are left in the field but bent double just below the cob to allow it to dry while ripening. Because the corn acts to suppress the spreading of weeds, further weed control is not necessary.

39

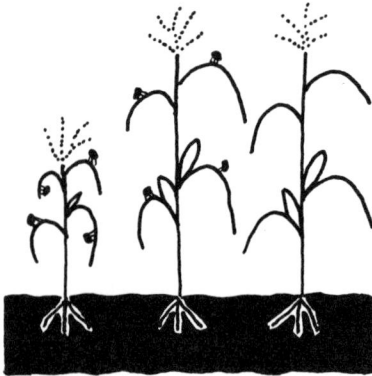

Figure 4.10. (a) Land races are quite variable varieties well-adapted to the region they are grown. They usually give a stable, low yield.

(b) In case a pest occurs, some plants may be resistant, some others may be tolerant or susceptible.

Local varieties

Varieties which are not reproduced and distributed centrally but which come from seeds collected through generations by the farmers theirselves, are usually to a degree horizontally resistant to pests common to that region. These varieties have been selected according to the principle of survival of the fittest. The varieties selected in this way are usually heterogeneous. In practical terms this means for instance that the one plant may be more resistant to one particular disease and another plant more resistant to another disease or perhaps to drought, so that risks for disasters are spread. The chance of a harvest totally failing is smaller than with genetically homogeneous, High Yielding Varieties. Nonetheless, pests can still cause heavy losses in local varieties.

Breeding for resistance

During the breeding process it is fairly difficult to build horizontal resistance into a new variety. Vertical resistance is easier to identify and to incorporate. Total resistance of new varieties is almost always vertical resistence.

If vertical resistance is broken by a pest, the search for resistance genes which can be crossed in into new varieties must begin all over again. In the meantime there is no resistance to the pest and usually chemical treatments must be resorted to. Sometimes, traditional varieties with a higher horizontal resistance are temporarily grown again. Sometimes further cultivation is almost impossible until a new resistant variety is distributed. Resistance breeding thus is a continual struggle against the adaptability of pests. On the other hand, efforts are done to breed varieties which have horizontal resistance to their mayor pests.

Centrally produced seed

The history of resistance breeding contains a warning for the farmer who does not reserve self produced seed. It is certainly true that in many situations and given sufficient artificial fertilizers, water and pesticides, the yield obtained from breeded, centrally produced and distributed varieties is higher, which makes the switch to a new variety attractive. If, however, the result is that self produced seed is no longer reserved, then there is no longer a way back and the outcome will be permanent dependence on the centrally produced, uniform varieties, which is quite risky. If circumstances should change, e.g. no rain, no pesticides or fertilizers available, these varieties often produce less than the traditional varieties under similar conditions.

Countless numbers of pests, including weeds, are spread via the seeds. For farmers who reserve their own seed it is therefore very important that only seed from healthy plants is gathered. There are, however, some pests which are barely noticeable in a field but which cause damage to a following crop. In this situation, centrally distributed seed is more reliable than self produced seed, because special care can be taken to ensure that proper hygiene is followed when the seed is produced. Its storage and eventuel treatment with pesticides is cheaper and safer than if this is done by the farmer.

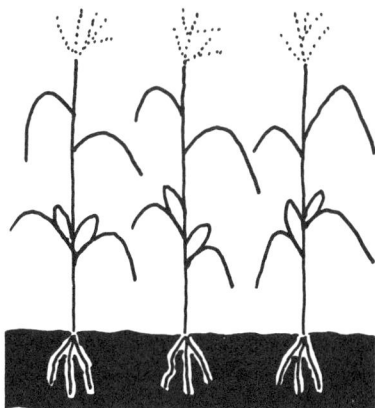

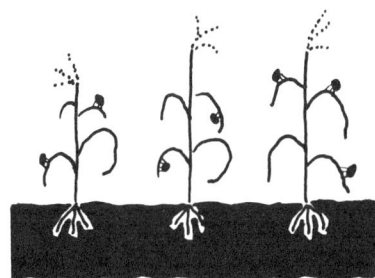

Figure 4.10. (c) .High yielding varieties are uniform crops. In optimal circumstances they give a high yield. (d) Under suboptimal circumstances, for example when a pest attacks the crop, great losses may occur because all plants will be affected.

Figure 4.11. In case crops are planted staggered, a pest population gets the chance to move from a harvested crop to a younger crop. In that case great damage can be done to the crops planted later.

4.3 Practices during and after sowing

4.3.1 Moment of sowing

In many tropical countries agriculture depends on the rain and the sowing date is set according to the start of the monsoon. Supplementary irrigation can then help to extend the growing-season in relatively small areas. Irrigation or a balanced rainfall may allow cultivation to be spread over a whole year. In certain seasons an individual pest may cause damage. A carefully selected sowing date can then be important to avoid this pest.

Example

In Brazil, the late dry season sowing prevents damage being caused to beans by the virus of the golden mosaic disease. The cooler temperatures cause a drop in the population of the transmitters of this disease, the white fly Bemisia tabaci.

Example

In India the low temperatures prevailing in late December give the peas an advantage over Fusarium solani, the cause of root rot.

Even when the sowing date is determined by the rainy season, a variation of a couple of weeks, whether earlier or later, can make a lot of difference. Many insects have a fixed life-cycle, predominantly determined by seasonal changes, such as the length of day and the onset of the rains. For instance, many stem borers only become active after the first rains have fallen. By carefully selecting the date of sowing, it is possible to avoid the most sensitive stage of the plants coinciding with the period when the density of a harmful insect population is at its greatest and when the possible risk of infestation is at its highest.

Example

The rice stem gall midge Orseolia oryzae can seriously damage the rice plant if the harm takes place before or during the tillering stage, after which rice is no longer a host for this insect. Early transplanting of the rice plants so that tillering takes place earlier in the season proved to be a solution in North Thailand.

Example

If cotton is sown later than usual, some of the moths of the pink boll worm Pectinophora gossypiella lay eggs while the fields are still void of cotton. The larvae will consequently die without doing damage.

41

The second important aspect of the sowing date for crop protection is the synchronised sowing of large areas. Many pests are directly connected to particular crops. If the sowing of a crop is spread over a number of months, a pest will be able to nest and develop in the fields that were sown first and inflict much heavier damage on the later crops. Synchronised sowing avoids this problem.

4.3.2 Depth of sowing and planting

Top soil prevents rapid dehydration of the seed and also protects it from being devoured by insects, birds and rodents. However, if the seed is sown too deeply little will rise above ground. The same applies to the planting of tubers and root stocks.

When transplanting seedlings from a seeding bed, the depth of planting is acutely relevant. If planting is too shallow, the seedlings will fall over or dry out. However, burying the collar of a plant makes it susceptible to infection from soil diseases.

4.3.3 Density of sowing and planting

As with all cultivation practices, a number of factors determine the density of sowing, the health of the crop being only one. The availability and cost of the seed, the water supply, the danger of erosion, etc. all play a part in this.

The density of planting influences ambiguously. In a more densely sown crop, the canopy closes more quickly and the crop is placed in a stronger position to compete with weeds. This often means that further weed control is not necessary. High plant density also helps to prevent many insect infestations. Insects are often attracted by fields where the individual plants can be seen. Densely planted crops form one uniform green surface, with the result that fewer insects settle in the field. The same applies to diseases which are transmitted by certain insects. A closed leaf surface is also beneficial when using the sprinkler irrigation method as it considerably reduces the risk of infection from mud splashes.

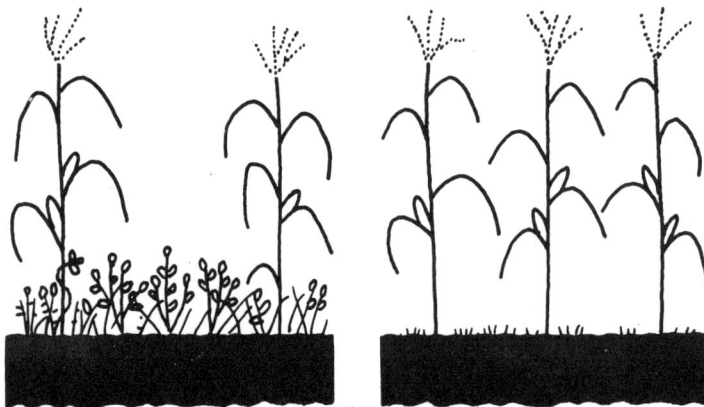

Figure 4.12ab. Widely planted crops allow weeds to grow between the plants. Therefore, for weed control it is better to plant a crop densely, so that it closes its canopy soon.

Example

*The groundnut aphid **Aphis cracci-vora** transmits the groundnut rosette virus and can consequently cause serious damage. If the ground-nuts are sown densely, the crop will quickly close up and the aphid which is attracted by semi-bare soil will be deterred. Losses incurred through the virus will therefore be reduced.*

Example

*The horn bill beetle **Orydes** is at-tracted by single standing coconut or palm oil trees and by palm trees which rise above the others. There-fore, synchronized planting at rela-tively close distances will ensure that a uniform green surface devel-ops which will not attract the beetle.*

Figure 4.13. Removing infected parts of plants inhibits certain pests from spreading over the crop. This measure is especially of importance for tree crops.

In the case of diseases which are transmitted directly from plant to plant, the situation is direct-ly opposite. High crop density facilitates the transmission of many diseases prevalent in shoots through direct or aerial contact between stems or leaves brushed against each other by the wind. The same is true for nematodes and fungi of roots because of the shorter distance between roots in the soil. Secondly, greater plant density brings about a different microclimate between the plants. The humididy then increases and this is favourable for many bacterial and fungal diseases.

If water is scarce, greater plant density causes more water stress and related diseases such as stem rot in corn.

Usually farmers know very well the optimum crop density. Experimenting with different plant densities with the intention of enhancing the control of weeds or other pests will only be meaningful if the water supply can be improved or if other limitations such as lack of seed or artificial fertilizers can be avoided. Then changing the plant density can indeed be a powerful practice to improve the health of a crop.

4.4 Practices during the growing season

4.4.1 Mulching

Mulching implies depositing or leaving a layer of plant remnants in the field. A mulch serves as a protection against erosion, influences the temperature of the soil, reduces the evaporation of water from the top-soil, improves the soil structure and encourages micro-biological activity. In many cases the damage caused by insects and diseases proved to be less in cases where a mulch had been deposited and there was also evidence of a strong reduction in the quantity of weeds.

Of course, care must be taken to ensure that a mulch does not become a source of weed seed or harmful insects. Therefore, remnants from the same crop or of a crop similar to the one in the field should not be used as mulch, the chances of creating a source of infection are great then.

Example

*It appears that the scale **Saissetia coffeae** and **Astrolecanium coffeae** and the coffee thrips **Diathro-thrips coffeae** inflict much less damage on coffee after a mulch has been deposited. Parasites of these in-sects are able to develop in this mulch.*

4.4.2 Sanitation

Already infected plants are an important source of infection for the crop. Removing such plants is an important means of controlling some pests. Once gathered, the contaminated plants or segments must be burnt, composted or buried. The practicability and the usefulness of such measures should, of course, be assessed for each situation. Tree crops in particular benefit from planned pruning and the removal of dead wood and fallen fruit.

Example

Subsistence farmers in Irian Jaya control taro blight (a fungal disease) simply through the daily removal of infected leaves.

Example

*If fruit trees are infected by fruit flies **Tephritidae**, all infected fruit must be plucked or collected from the ground to be destroyed in order to avoid spreading of the infection.*

Example

*To control the coffee berry borer **Hypothenemus hampei** all the old, ripe and dried out berries should be removed from the bush. If this is done efficiently, further control is not necessary.*

Example

*As part of the control of the stag beetle **Oryctes spp.**, a pest found in oil palms in Asia, the fallen palm leaves are cleared away and dead trees cut down and burnt.*

Apart from competeting for food, light and water, weeds can also indirectly damage a crop if they serve as alternative host plants for pests. Such weeds must be kept out of the vicinity of the field or they should be removed by weeding.

Example

*Many pests which cause injury to rice use grasses as alternative hosts, as with the green rice leafhopper **Nephotettix nigropictus**, the rice leaffolder **Anaphalocrocis medinalis**, rice armyworms and the rice stem gall midge **Orycolia oryzae**. The destruction of grassy weeds in the vicinity of the rice fields is an important measure against these insects.*

Figure 4.14. A scarecrow may keep away birds for a short time, for example from seed beds during the days before germination.

4.4.3 Miscellaneous

Apart from employing the various cultivation methods already described, farmers have proved to be very inventive in deterring or driving away insects and birds. A few examples of this are:
- Nets, ribbons, cloth or plastic are stretched out in fields to prevent the insects or birds getting at the crops;
- Aluminium strips, unwound cassettes or tin cans filled with stones hung up above a field frighten birds away through the noise they make in the wind and their glitter in the sun. Some farmers try to keep birds away by hanging up the corpses of dead birds.
- Insects are kept at bay with smoke from burning car tyres.

Figure 4.15. In many regions traditional rites ensure a good yield. Priests bless the crops, Gods may be offered sacrifices for rain, etc.

Wind breaks, are first of all planted against wind to protect the crop for storms. So they also act as a brake on the spreading of many wind carried pests. The reduced wind speed behind the wind tunnels acts as a trap for spores. Because of this, a space should be left between the wind tunnel and the field. Some fungi and bacteria take advantage of leaf damage caused by the wind. By reducing the chance for such damage, wind breaks also offer protection against these pests.

44

Example

In the Philippines bacterial diseases in rice add greatly to losses left after a tropical cyclone has passed. This is attributed to the amount of slight leaf damage which occurs in strong winds.

Example

The cotton muzzled beetle Anthonomus grandis is controlled by starting the cotton cultivation as early in the season as possible. Early sowing, the use of early varieties and a quick harvest are all parts of the control mechanism.

Example

Many stem borers (family Noctuidae, Pyralidae) pupate in the lower part of the stems of crops as corn and sorghum for instance. During harvesting a length of the stem is usually left in the field (stubble). This enables the stem borers to sur-

In many regions, religious customs play a role in crop protection. In Buddhist regions, for example, mantras are sung to protect the crop against pests, storms and drought. In many Christian regions it is the custom to have a field blessed by a clergyman. These rites are often an essential part of a farmer's daily life.

4.5 Measures during and after harvest

4.5.1 Timing the harvest

As with the timing of sowing, the harvest period can be similarly used to escape from the peak period of certain pests.

Most pests multiply as the season progresses. This means that the harvest should take place as early as possible. Other factors, however, also play a role: cereals and pulses for example must be allowed to mature fully. A quick harvest under favourable conditions is important as infestation just before the harvest can result in an epidemic when the harvest product is stored. Chapter 7 of this part discusses storage problems more in detail.

4.5.2 Disposal of plant remnants

Many pests survive in the remains of a crop left on the field. If the plant remnants are destroyed, the level of the pest organism population at the start of the next growing season will be much lower. The remains can be burnt, grazed or ploughed under. Even making sure that the remains are spread out flat on the soil will help, as rotting will then occur much more quickly.

4.6 Conclusion

Traditional cultivation systems especially are often primarily aimed at achieving a balance between crop, pests and their natural enemies, so that serious losses at harvest time are avoided. Cultivation practices which are intended to provide better pest control are in general environmental sound and not expensive. Their effectiveness, however, cannot always be predicted, and they can have undesirable side-effects. For this reason, alterations in cultivation practices should always undergo small-scale tests during a number of seasons and under farming conditions.

Moreover, because cultivation measures usually affect different aspects of cultivation, they can only be properly devised after on-farm experiments. Solutions which have been applied elsewhere may determine the general direction to be taken but they can rarely simply be copied.

5. Biological control

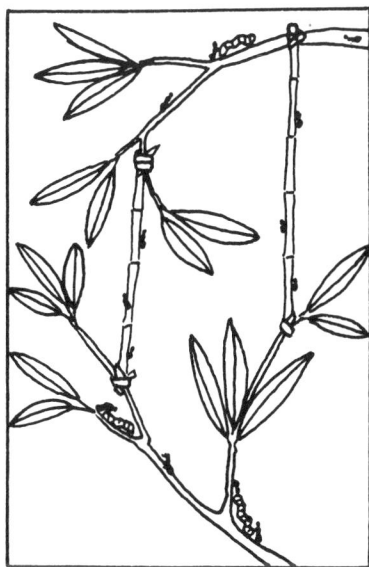

Figure 5.1. Helping natural enemies: ants predate on caterpillars in fruit trees. When branches are connected the ants are in a position to control the pest completely.

There is hardly a pest to be found which is not naturally opposed by other organisms. Such organisms, known as "natural enemies", can be pathogenic viruses, bacteria or fungi, parasitic insects (often ichneumons), predatory insects (e.g. ladybirds), spiders, birds or fish. Natural enemies of certain pests are present in every crop and most potential pests are therefore suppressed before they can cause any real damage. In principle, it is possible to help present natural enemies in controlling pests; many of the cultivation practices for example work by creating a climate favourable to natural enemies.

In the dry and cold seasons natural enemy populations are reduced, so that massive numbers of natural enemies sometimes come too late to prevent the pest from exceeding the damage threshold. The effects of chemical control can be another reason for a pest developing faster than its natural enemy. Broad-spectrum pesticides do not only kill the pest, but also its natural enemy. The frequent result is that the pest rapidly recovers, a new pest appears ("man-made pests"). If chemical control had not been used these pests would have been prevented from reaching dangerous levels by their natural enemies. This is a common situation and is one of the major disadvantages of chemical control.

Example

In the rice fields in the Chinese province Hunan, the following program was used to encourage predators, and spiders in particular, as a means of controlling cicadas:
- *cultivating winter crops to provide the spiders overwintering sites;*
- *transferring the spider's eggs to new rice fields;*
- *collecting spiders by placing bundles of straw in the rice fields during irrigation, and providing shelter during harvesting, e.g. by planting soya along the edges of the field;*
- *selective use of insecticides (see chapter 6).*

These methods greatly reduced the use of insecticides and in some years made them completely superfluous.

Example

*The caterpillar **Anticarsia gemmatalis** is the main pest of soya in Brazil. As means of control the farmers introduce caterpillars which are infected with a certain viral disease. At the end of the season the diseased caterpillars are collected and re-introduced at the beginning of the following season. This measure greatly reduces need for the use of insecticides.*

47

Figure 5.2. Leaving a ground cover of grass in vineyards in California results in a habitat modification that enhances the activities of predators of phytophagous mites.

The biological control of insects and mites

There are three types of natural enemies: parasites, predators and pathogens. All three types are used for the control of insects.

Predators are mostly insects, which in one life-time will consume a number of other insects. They actively seek their food, normally kill their prey and usually have a life-span longer than that of their prey. Examples of predators are preying mites, ladybirds and lacewings.

Parasites are in general specialized insects or mites which lay their eggs in, on or close by individuals of another species. The larvae live on the blood and the tissue of the host and do not need actively look for their food. On reaching adult stage, they are no longer parasitic but fly about looking for a new host for their offspring. In its lifetime an individual parasite consumes only one host, while females often lay a number of eggs. Examples of parasites are ichneumons, i.e. *Encarsia formosa, Trichogamma* spp. which are applied commercially in greenhouses.

Pathogens (parasitic microorganisms) cause their hosts to sicken and to die. After death, millions of individual microbes (spores) are released which disperse easily on the wind and/or in water. Because of their minute size and ability to reproduce, Pathogens lend themselves to mass-breeding and can be spread over a field with spraying equipment. An example of a pathogen is *Bacillus thuringiensis*. This bacteria is already frequently applied, and produces a chemical which is toxic for insects, but which hardly affects plants and vertebrates. This success can be credited to the fact that it is commercially competitive with pesticides. Preparations of *B. thuringiensis* can be produced by small industries, as now happens in China. Some viruses, fungi and protozoa can be used for the biological control of pests as well.

Artificial biological control can be applied either by repeatedly introducing natural enemies (5.2) or by taking a natural enemy from elsewhere and leaving it to survive in the new area (5.3). The chapter ends by two chapters on the sterile male technique and on the use of chemicals produced by insects itself.

5.2 Repeated introduction of natural enemies

For repeated introduction of natural enemies a great number of a natural enemy is bred in the laboratory (mass-rearing) and released in the field early in the season. This method uses the specific action of a natural enemy to keep the pest below the damage threshold. The greatest number of successes in controlling pests has been achieved with the introduction of parasitic insects (especially ichneumons), predators (insects, prowling mites, spiders), nematodes and insect viruses. Many water weeds have also been successfully controlled by using fishes and insects. The use of antagonists against diseases has until now not been very satisfactory; only incidental success has been credited against vectors of diseases and against nematodes.

The regular release of natural enemies requires a mass rearing and distribution system, backed-up by good information facilities for users. As do many forms of pest control, this entails constantly re-occurring costs. If the natural enemy is a virus or a fungus which attacks harmful insects or plants, in many cases it can be distributed by spraying viruses or fungal spores in the same way as pesticides. Apart from the very plausibility of introducing specific natural enemies as pest control, other criteria which should be given prior consideration are the persistency of the relevant vector(s) and the possible hazards for health and the environment.

Because biological control is as a rule very specific, successful control of one pest, the attacked pest may be replaced by a competitive pest which is resistant to the treatment. As crops usually harbour numerous types of weeds, this problem can be especially observed in weed-control. Rice-borers also, do not only damage the crop but compete amongst each other; treating one type, even if it is the most common, can give unsatisfactory results.

5.3 Importing exotic natural enemies

Figure 5.3. Interior biological control. Spiders are very useful to control irritating insects like flies.

Crops often originate in another region then where they are cultivated. These "imported" crops are often relatively free of pests and disease as the causes of such injury were left behind. Recent attempts have been made to shut-out foreign pests and diseases by following strict quarantine measures when importing new plant material. Nonetheless, pests have been and still are being carried to all parts of the world. If the natural enemy of such a pest is left behind, it will multiply unhampered by its antagonist. If an important crop is concerned, it is worthwhile returning to the original area to find natural enemies which can be introduced in the new area.

Before a natural enemy of a pest is introduced into a new region, lengthy and expensive research has to be conducted. In the first instance, this research must find natural enemies of the pest in the area of origin and what organisms the natural enemy attack. A natural enemy which also attacks useful organisms must not be imported to another area, it may do more damage than have positive effects.

Once a natural enemy has been tested extensively and is successfully introduced in one region of the world, applying it in other countries is considerably easier.

Example
There has as yet been no success in controlling grassy weeds by using insects. Insects which caused sufficient damage to the grass consistently appeared to be too dangerous to related cereals.

Figure 5.4. (a) A predatory mite attacking another mite.
(b) A parasitic wasp laying an egg in an aphid, which will be eaten by the larva of the ichneumon.
(c) Ladybirds are wolverines, attacking all kinds of insects.
(d) Ladybird's larvae are less known, but also play an important role in suppressing pests.

Example

*The water hyacinth from South America has become a major weed in lakes all over the world. Its numbers are successfully controlled by the introduction of two natural enemies: a weevil **Neochetina eichhorniae** and a mite **Orthogalumna terebrantis**. After extensive investigations, these have already been introduced into a number of countries. Subsequently, their further distribution has become very simple and without risk.*

Ideally, the imported natural enemy will then be able to survive without outside assistance and multiply and spread out with sufficient speed. In that case a single introduction of the antagonist suffices.

Example

*The scale **Icerya purchasi** was accidentally introduced from Australia to California in the last century and caused major damage to citrus crops. To fight the pest, the ladybird Rodolia cardinalis was introduced in 1890, succeeded in permanently establishing itself, and still acts as an effective control of the scale.*

At this point, however, a warning must be sounded. Without expert assistance no attempt should be made to introduce natural enemies into a region. There is a grave risk of also introducing undesirable organisms such as parasites of the natural enemy, or a disease of the crop to be protected.

5.4 Sterile male technique

The sterile male technique is applied exclusively for the control of insect infestations and can only be used in a limited number of situations. For this technique a large number of males has to be bred and sterilized. At a well chosen moment these males are released in such great numbers that the existing fertile males are given almost no chance to mate with the females. As a consequence, a large percentage of the females mate with the sterile males and, therefore, do not reproduce. In certain cases this technique proved to be so effective that the relevant plague has been totally rooted out of the region where it was applied. For example the extermination of the tse-tse fly in large areas of Africa is done by the sterile male technique.
The technique demands extensive preparatory research and as well as a breeding and distribution system. Nonetheless, the costs in both financial and environmental terms are small in comparison to those of spraying pesticides.

50

Figure 5.4. (e) Besides spiders spinning webs, there are a lot of species which chase their prey. Of some spiders in rice it is known they eat about seven brown planthoppers a day.
(f) If frogs are allowed they eat huge amounts of insects.
(g) Many birds eat caterpillars or snails.
(h) Cats, dogs and snakes chase rats and mice, and may be very useful, especially in combination with destroying hiding places of the pests.

5.5 The use of hormones and feromones

Hormones. As in humans, some biochemical processes in the body of insects are regulated by hormones. By spraying these hormones as insecticides, it is possible to disturb the development of the insects which they contact. A number of chemical combinations have been recognised which act in the same way as hormones but which can be artificially produced. The effect of applying hormones to control insects are similar to effects of pesticides.

Feromones are chemicals working totally different as they act outside the body of the insect. They are dispersed by the insect in the form of scent which acts to transmit a particular message. There are sexferomones which are excreted by the females to attract the males; aggregation and distribution feromones which regulate the number of individuals per plant or per surface area; and alarm feromones which sound an alarm signal.

Feromones are specific for every species, each insect uniquely recognising its own feromones. Very low concentrations are required for them to be effective.

Feromones are often used in 'feromone falls', for example to keep an eye on the population level of an insect infestation. In principle, feromones can be used to disturb an insect population and due to their specific characteristic they are particularly suitable for use in integrated pest management. Their use, however, is still in its infancy. Successes have mainly been achieved with sexferomones.

Example

The pink boll-worm **Pectinophora gossypiella** *has been effectively controlled in many regions by ensuring that the air between the cotton plants is totally saturated with the sex feromone of this particular insect. The feromone is applied to the plant by means of a 'slow-release' packaging system and prevents the males from finding the females.*

Figure 5.5. A feromone trap attracting a lot of male butterflies. The poor females are not able to attract males no more, their eggs are not fertilized and there will be no offspring to do damage to the fruits.

5.6 Final comments

In practice the main form of biological control is natural control. This means for the farmer, that his or her action should lead to conservation of the natural enemies. For this pesticides, in particular insecticides, should be applied selectively. A main step in this is making the farmer recognize natural enemies.

Biological control can offer a high level of pest control at low costs, is environmentally sound, safe, directs itself specifically at the organism to be controlled and rarely leads to resistance. The reason why not much more is being done to introduce biological control is the costs of the necessary research and of extension required to apply such control. It is not that the costs of biological control are higher than those of chemical control; they are often even lower. The problem is that biological control is, apart from a few exceptions, often difficult to commercialise. Natural enemies cannot be patented, which makes the investment of capital into research into biological control unattractive to industry. Consequently, the development of these methods heavily depends once again on financial backing by government and on pressure from international organizations.

6. Chemical control as a part of IPM

This chapter explains how pesticides can be used as efficiently as possible as a complement to other pest control methods.

6.1 Pest populations and chemical control

As Chapter 3 pointed out, by no means do all pests cause losses. The previous chapters discussed the various ways of keeping pest levels below the damage threshold through cultural practices and natural enemies. If, after taking these measures, the pest population still threatens to exceed the damage threshold, then curative pest control measures must be considered. This chapter deals with the "emergency measures" available against the development of a pest population: chemical pest control. Chemical control is by far the most important curative measure. Other pest control techniques include hoeing and weeding, or in rare instances, the repeated introduction of natural enemies.

6.1.1 The deliberate use of pesticides in theory
Properly, all the advantages and disadvantages of a technique should be weighed against each other – a process which implicitly often is carried out by the farmers. This also applies to the use of pesticides. Rationalizing the use of pesticides is an important step toward reducing their use, and basically means that spraying only takes place if it is thought that the cost of pesticides will be lower than the expected increase in yield. Spraying is carried out in the expectation that as a result the yield will increase. In theory, the costs of spraying can be weighed against the benefits of an increased yield. The damage threshold is therefore at a level where the costs of spraying and the benefits from the extra yield are more or less equal.

Because it may take some time before a field can be sprayed and for this to become effective once a pest has been detected, an economic threshold has to be established. The economic threshold is the level of pest population at which control can prevent it from exceeding the damage threshold, this is shown in figure 6.1.

6.1.2 Problems rising when applying thresholds
Determining the damage threshold is not easy. Both the effect of a treatment and the future price of an agricultural product must be known. Extensive field- and on-farm testing should be carried out. For such tests a comparison is made between two fields as identical as possible, one of which is infected, the other "clean". Because this sort of testing is expensive and separate tests have to be carried out in each area as thresholds vary considerably from region to region, official damage thresholds are not widely available. The economic threshold level is sometimes reached before it can even be observed.

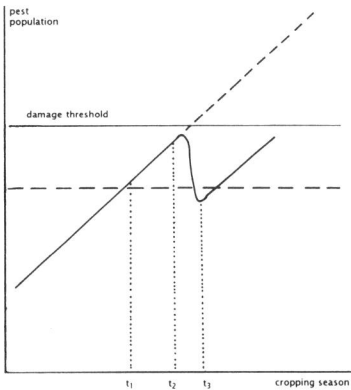

Figure 6.1. The farmer observes his crop, and notes that the level of a particular pest population has gone beyond the economic threshold (t1). Spraying cannot take place until the following morning as time is needed to prepare the apparatus (t2). It takes some time before the pesticide takes effect, but the pest population is subsequently reduced to below the economic threshold and damage to the crop is avoided (t3). In the heat of the tropics a pest population can multiply so rapidly that in one growing season repeated sprayings may be necessary.

Disadvantages and advantages of pesticides

Disadvantages of the use of pesticides are:
- pesticides require large investments of farmers;
- skill is needed to apply pesticides in a proper way, they may be hazardous to the users;
- one application of pesticides often necessitates more applications because of a quick recovery of the pest population;
- natural enemies of insect pests may be affected which results and resurgence of (sometimes even unknown) pest insects, so-called "man-made pests";
- ever higher dosages may be required to get the same result eventually leading to resistance of a certain pest against a pesticide;
- they cause damage to a lot of organisms, directly (e.g. birds eating treated seeds, or bees) and indirectly (by the accumulation of persistent pesticides as organochlorines in the food chain);
- they pollute sources of drinking water;
- residues of pesticides contaminate food;
- and last but not least: the production of pesticides is very hazardous for man and environment. In the case of a catastrophe in a pesticide factory the consequences are disastrous (e.g. Bophal and Seveso), but also when no catastrophes occur the production of pesticides is very polluting and causes health risks for people living in the environment.

The disadvantages apply to negative effects on flora and fauna, on quality of soil water and surface water, on agricultural sustainability (pollution of the soil and eradication of useful organisms), and on public health.

Advantages of the use of pesticides.
Why are pesticides used so much in spite of all these negative effects?

There are a number of reasons:
- pesticides can be used directly against a certain pest, which suddenly becomes epidemic, as a sort of emergency-brake; often application of a pesticide is the only way of protecting the crop and thus of preventing yield losses;
- they replace labour-intensive work like weeding;
- a very important reason is that, while the disadvantages affect the society as a whole, the profits go to small groups of people, first of all to the producers but also to farmers.

If cultural measures, such as cultivating resistant strains, maximum fertilization etc. are not effective and there is no alternative remunerative crop, then the possibility of carrying out calendar spraying must be considered. This is particularly the case in a number of fungal diseases which due to the heat and the moist atmosphere in the tropics can spread with lightning speed. This measure, however, cannot be considered as IPM and will be given no further detailed consideration.

Unfortunately, calendar sprayings are carried out far more often than is actually necessary. The reason for this is that most farmers in the tropics are virtually untrained in recognizing pests. The farmer wants to take as few risks as possible and consequently sprays after the first observation of what may possibly be an innocent beetle and not a pest at all. In practice therefore, the economic threshold is not only influenced by the crop, the pest and the regional prices but also by the ability of the farmer to observe and to judge.

6.1.3 Pest thresholds in practice

Economic and damage thresholds are, unfortunately, rarely known and therefore other methods have usually to be followed. If farmers recognise a pest, their many years' experience with the crops have taught them at which population level and at what stage of cultivation a pest can cause damage. The information possessed by the farmers is often found to be very close to the values established later after scientific investigation. Farmers, however, often

Figure 6.2. Farmers should be tought to recognize pest insects and their natural enemies. Only when they know which organisms are pests, they may be able to judge whether spraying is necessary or not.

mistrust their own thresholds, either because of advice from the local pesticide dealer, or sometimes because of the regional extension centre.

If farmers do not recognise a pest and see every creature or every mutation as a potential danger to their crop, then learning to identify pests must be the first step. If a farmer can distinguish the pests from harmless organisms or other symptoms in his of her crop, a damage threshold should be determined. This is done in the following manner, preferably by a few farmers acting simultaneously. The nature of the damage that a pest is inflicting on a crop is closely examined. If the damage is considerable, an estimate of the damage to the harvest is made in cooperation with the farmers. If this appears to balance out against the costs of pest control then treatment will take place; otherwise not. By applying this method in different fields with different groups of farmers, a damage threshold is obtained which the farmers trust. Apparently farmers do trust results if these are mutually confirmed. What is involved here is not the establishment of an exact damage threshold, but the fact that damage thresholds are brought into use at all - even if these are slightly on the low side. Farmers should be brought to view spraying as an investment with both good and bad effects and no longer as a sort of insurance against crop loss.

For an example of a crop protection programme working in this way see paragraph 12.4.

To give an idea of the variation level of damage thresholds, the damage thresholds for a number of pests, diseases and weeds in maize and cotton are given below. In general, little distinction is made between damage and economic thresholds. The thresholds given here for one region, even differ from place to place and from year to year. It is, therefore, impossible to transfer them gratuitously; however, they do give an indication of the level at which consideration should be given.

55

Damage thresholds of cotton in Turkey

In the Kilikien Plain, South-East Turkey, cotton is grown as an important cash-crop. Since the 1960's cotton yields have increased rapidly due to the improvement in agricultural practices such as irrigation, fertilization, drainage and mechanization. Crop protection remains a major problem. Because of the climate, diseases rarely occur, but pests are of immense importance. Cotton is grown from April till October in Turkey. *Bemisia tabaci* generally occurs from the end of May to the end of the season. The larvae and adults suck sap from the leaves and cause serious leaf fall. The production of honeydew reduces photosynthesis by the growth of fungi on it. At higher population densities, the underside of the leaves are covered with wax excretions and coatings which plug the stomata, and plants will die rapidly. The entire life cycle requires about 10 days during the summer.

Control measures are necessary when more than two adults per leaf are found. This is estimated by taking a sample of at least 100 leaves from 20 plants: two from the top, one from the middle and two from the bottom of the plant.

Heliothis armigera is a night-moth. Damage becomes apparent in the first week of June. The larvae feed on young leaves and, in particular, bore into buds. These buds open as usual, become yellow with time and drop. At the beginning of September, the larvae bore into the bolls and remove the entire contents. One generation takes 3 to 6 weeks and there may be two or three generations a season.

An average of more than 0.5 lava per plant is economically damaging. A sample of 3 x 3 m rows (= 3 x 12 plants) has to be taken.

Spodoptera littoralis usually becomes abundant in the second half of July. During the day, the larvae are found in the soil near the root system of the plants. During the night, they crawl up the plant and feed on the leaves or above-ground plant parts. At first, many small feeding sites are present which then expand over the whole leaf area. At high population densities, the whole plant is damaged within a short period. The larvae, in particular the 2nd-4th stage ones, consume large quantities and so severe damage can occur. Plants are attacked almost up to harvest time. Three or four generations may be produced per season.

At least three samples of 6m rows have to be taken (= 24 plants per row). When more than one larva per plant is found, control is necessary.

By applying the pest management described above, the average number of pesticide applications could be reduced from ten to five per growing season. This is also conducive to the conservation of many natural enemies, although in order to achieve this result, the growers must be thoroughly trained.

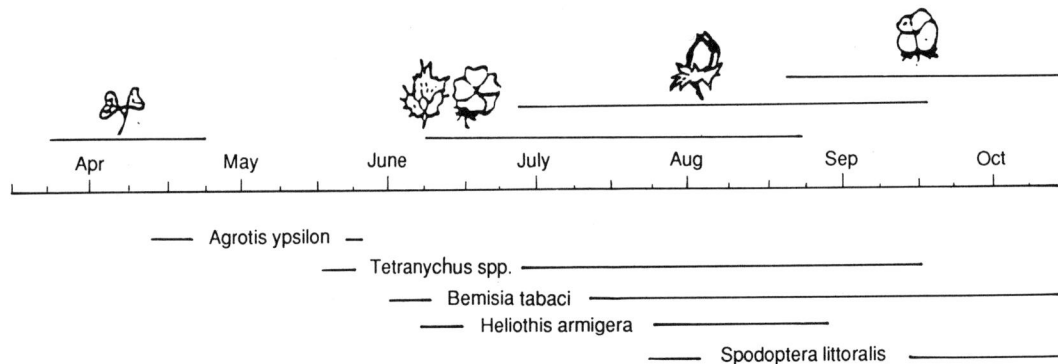

Figure 6.3. The major cotton pests in Turkey and their time of appearance.

The *maize leaf aphid*, a soft, bluish-green aphid about 3 mm long, gives the upper leaves a brown and sticky appearance. In the USA the economic threshold was established at being 5 adults per plant 3 weeks before tasseling.
The *fall army-worm* tears leaves to pieces or eats away at the leaf-margins or makes holes. Later kernels are also burrowed out and eaten. The worms are dark green and up to 2 inches in length with lighter stripes on their sides and down the middle of their backs. The worms feed during the night and shelter during the day. They often enter the field in large numbers from nearby grasses or small grains. In the USA the economic threshold was established at 0.2 worms per plant in the early whorl stage, 0,6 larvae per plant in the middle whorl stage and at 0.8 larvae per plant in the early tassel stage. In Nicaragua, control measures are taken when 20 per cent of the whorls are infested during the first 20 days of plant development and at tassel emergence.

6.2 Effects of pesticides

Before going further into how pesticides can be most efficiently applied (6.5) an explanation will be given of the side-effects and the dangers of their use followed by some advice on how they can be safely managed.

Pesticides are usually divided into their respective target groups, in other words, they are categorised according to the organism they have been produced to attack. The most important of these groups are: insecticides (against insects), herbicides (against weeds), fungicides (against fungi). Other groups are: acaricides (against mites); nematicides (against nematodes); rodenticides (against rodents – especially rats); bactericides; algicides; molluscicides (against snails); piscicides (against fish) and avicides (against birds). Almost anything which blossoms and grows can be poisoned, the only exception being virus diseases against which there is, as yet, no counterattack.

Pesticides can also be named and divided according to their chemical structure. The naming of a chemical can be somewhat complicated. The chemical name of the herbicide Paraquat for example is 1.1-dimethyl-4.4-bipyridinium-dichloride, but the term "Paraquat" has become the name by which this chemical is internationally recognised. It is also the name which both industrial and scientific journals use to indicate this particular chemical. "Paraquat" is consequently the name used in the list of pesticides found in appendix I. If everything is as it should be, this will also be the name to appear on the label. If more than one international name appears on the label, then the contents are made up of a mixture of different pesticides; the names on the label bear direct relation to the active chemicals contained.

The name on the label written in the largest letters is the trade-name. The labels of chemicals which are marketed by one company only will bear one trade-name; chemicals which are marketed by more than one company will be given their own trade-name by each firm. Trade-names for Paraquat could for instance be: Dextrone, Gramaxone, Weedol and Esgram or perhaps also: Agrichem Paraquat, Luxan Paraquat, etc. In developing countries farmers are often more familiar with the trade-names than with the internationally accepted chemical titles.

The trade-names can sometimes seem rather ridiculous, especially if they have been adapted to the local colour. For example, you might come across "Ambush", "Roundup", or "Macho". However, these names have little significance for the discerning sprayer or extension officer.

Figure 6.4. (a) Broad-spectrum insecticides kill all kinds of insects, including harmless or even predatory ones. Therefore they have a wide range of application and they can be sold in huge quantities, which makes them relatively cheap.
(b) Selective insecticides, on the other hand, kill only few species. Their application is restricted to the few insects they act upon, so that they can be sold only in small quantities, which makes them expensive.
(c)

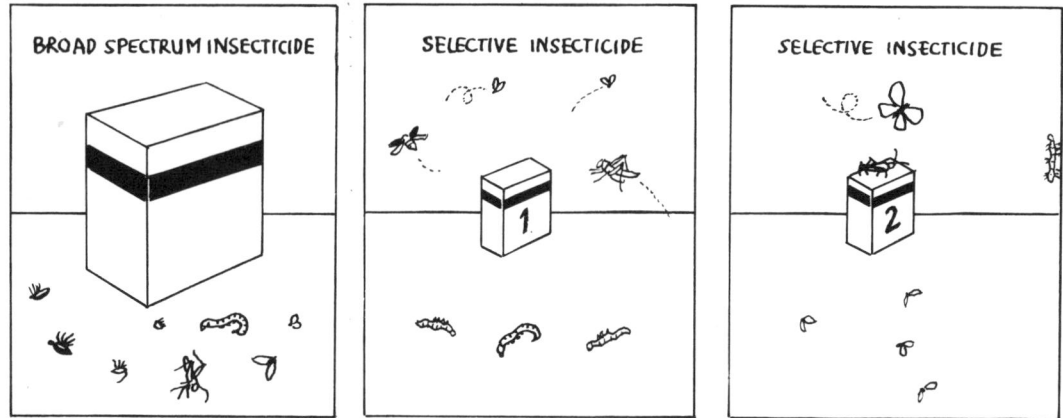

6.2.1 Effects of pesticides within agricultural systems

Selectivity. Apart from affecting pest organisms, a pesticide always damages harmless or useful organisms in the agricultural ecosystem. The amount of damage done is determined by the degree of **selectivity** of a chemical. A selective chemical kills a number of specific pest organisms but spares the crop and many other organisms. This is due either to a specific toxic mechanism possessed by the chemical or to the way in which a poison is used. Many chemicals, however, are non-specific in their action: many organisms are sensitive to their effects. They also kill off natural enemies and other useful creatures such as bees. Important pest control programmes have gone wrong because a too liberal use was made of non-specific chemicals causing the disappearance of almost all natural enemies of the pest. Sometimes pests could then develop (secondary plagues or "man-made pests"). In an IPM programme, selective chemicals or applications should play an important role; however, selective pesticides are still not available for many types of pests. Where a selective pesticide is available, it is often expensive – which has a strong influence especially on the small-scale farmer.

From a commercial viewpoint, the development of selective chemicals is of little interest. The costs of research into selective chemicals is the same as that for non-selective chemicals. But as a selective chemical only works against one type of pest, sales are less than for non-selective chemicals, which increases the price per kilo. Cheap non-selective chemicals dominate the market when it comes to world-wide use of crop protection products.

In the control of weeds, selective chemicals are a necessity; if a non-selective weed-killer were used, the entire crop would be wiped out. The method of applying a chemical plays an important role in its selectivity. By placing a shield over the spraying apparatus the spray can be restricted to reaching **between** the rows.

Figure 6.5. Selective methods of applying pesticides may increase selectivity of its action. Here a shield is placed on the lance, so that the crop is not affected by the chemical.

Resistance develops as the result of frequent spraying of the same pesticide. The population of pest organisms often consist of many millions. Each generation will always contain one or two individuals with mutations in their genetic material that render them less sensitive to the

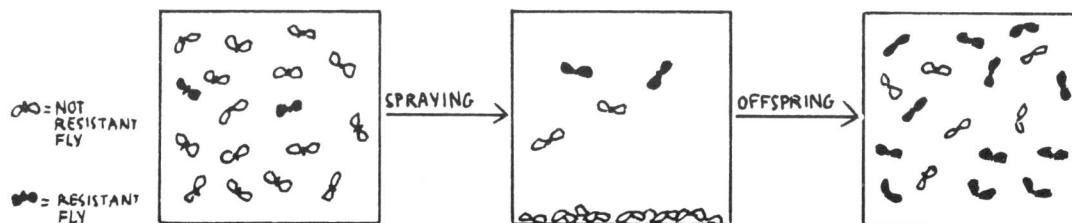

Figure 6.6. Development of resistance to a pesticide. Out of a huge population, a few individuals are resistant against the pesticide. These survive spraying and multiply, their offspring is resistant to the pesticide which loses its effect.

poison. Reproduction is often so rapid that new characteristics with a greater chance of survival become dominant (resistance).

There are many examples of the appearance of resistant pest organisms, fungal diseases and weeds. The first case of resistance was observed in 1947: a house-fly exhibited resistance to DDT.

The development of resistance can be slowed down by changing pesticide per treatment.

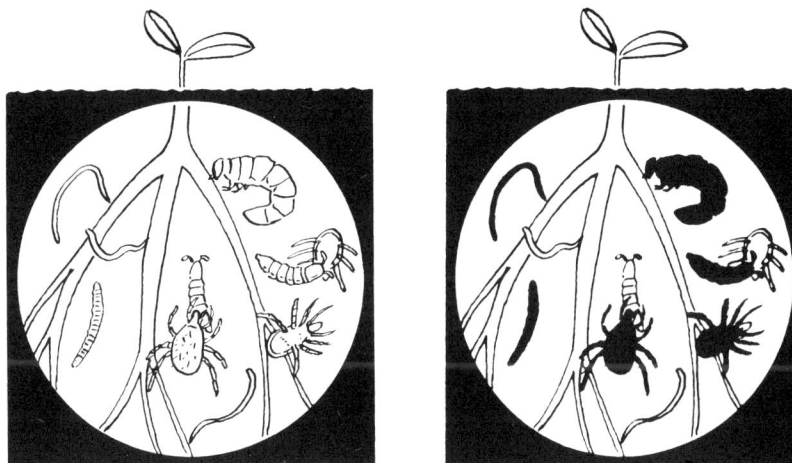

Figure 6.7. When applying a soil disinfectant, all kinds of organisms are killed. If afterwards a pathogen is introduced the natural antagonism will be diminished, and the pest will get the chance to cause a lot of damage.

Example

Atrazine is a herbicide that lends itself perfectly for use in maize crops as the maize breaks down Atrazine and is thus insensitive to its effects. However, if Atrazine is used repeatedly on the same field, the residue in the soil can reach such levels that any crop other than maize cannot be cultivated because of its sensitivity to the pesticide.

Effects on the soil. The frequent use of chemical pest control chemicals can damage the soil's ecosystem. The soil teems with microscopic organisms such as nematodes, spring tails, mites, fungi and bacteria. Little is known of this flora and fauna, how they interact and restrain each other. There is substantial uncertainty regarding the influence which pesticides have on flora and fauna. There is always the risk that useful organisms will be killed together with those causing damage, thus rendering the counter-effect unproductive.

Apart from their restraining effect on damaging organisms, useful organisms are also crucial to the conversion processes in the soil (e.g. the mineralization process). Sterilising the soil by means of non-specific chemicals destroys a great part of the flora and fauna. It is far better to treat the damaging organisms in the soil with structural methods such as crop rotation and resistant strains. Residues in the soil, in particular of herbicides, can possibly prevent the cultivation of some crops.

6.2.2 The toxicity of pesticides

Crop protection chemicals are intended to eliminate living organisms. This section takes a more detailed look at the subject of the toxicity of chemicals; the main purpose being to provide some background for the data in the list of pesticides in appendix I. This list comprises a number of frequently used chemicals, but it must not be thought that an unlisted chemical is not poisonous; we have merely given a selection of those most commonly in use. If you require a more comprehensive list, there are various organizations to which you can refer (for adresses see appendix III).

The most important dangers of pesticides are:
- **the toxic effect on people and animals** through direct contact with the poison during crop spraying. The toxicity of the residue - the final remains of pesticides on the treated crops - also represents a risk to the consumer in the case of food-crops.
- **the toxic effect of pesticides in the environment**, viz. in the soil, the water and in plants, and through possible accumulation in the food-chain.

The risks involved in the different chemicals vary widely: some chemicals are highly toxic on direct contact, especially for the farmers who use them, but break down rapidly in the environment, while other compounds are not poisonous on direct contact, but harmful through accumulation in the environment and in the food-chain and pose a threat to the consumer.

In the developed countries, most of the persistent pesticides (which accumulate in the environment and in the food-chain) such as organochlorines, i.e. DDT, are prohibited and have been replaced by less persistent chemicals i.e. organophosphorous compounds or carbamates. However, these chemicals are usually much more toxic in direct contact with humans.
We have mentioned merely a few groups of chemicals with regard to their persistence. However it is impossible to describe toxicity or persistence for each chemical group: the differences between the various chemicals are too great.

A well-known example is the insecticide DDT. When this chemical was on the market, its use was widespread. In the immediate post-war years, people smeared DDT liberally on their scalps to get rid of annoying head-lice. It was not until some years later that birds of prey, which occupy one end of the food-chain, evinced failure in reproduction. Through the massive accumulation of DDT in the food-chain their egg-shells had become much too thin and broke by themselves.

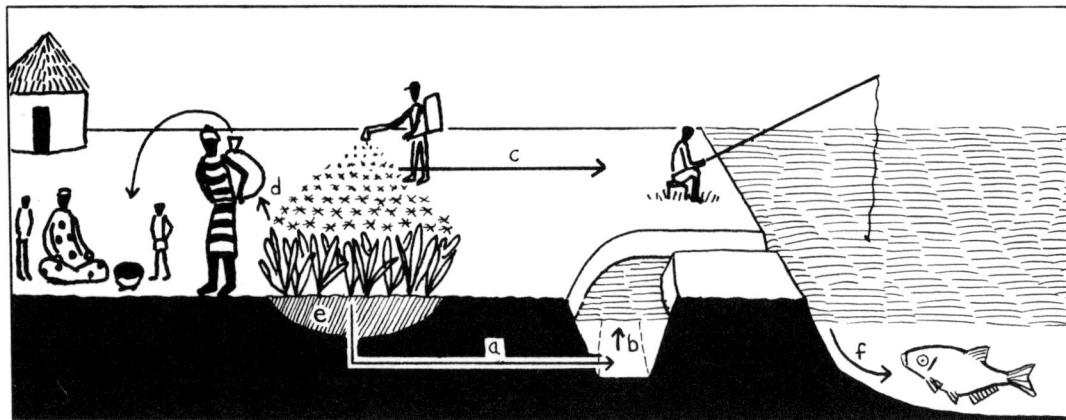

Figure 6.8. A pesticide is applied to a crop. Where will it end up? Via soil water it reaches surface water (a), and in the end the sea (b). The pesticide is blown away by the wind, or it evaporates from the crop and will settle elsewhere (c). The harvested product contains residues, by which it will get into food for humans (e). The pesticide stores in the soil and will decompose after some time (f). Via the food chain it ends up in fish or other organisms (f).

Effects on the environment
The environmental effects of pesticides are unavoidable: once applied they disperse into the environment. The quantity of a pesticide dispersing into the environment depends on the

manner of its application, the weather conditions and its character. Spraying from aircraft, in particular, involves a considerable loss of pesticides, up to 60–70%. Also with more usual application the loss can be quite substantial. Windy conditions during spraying provide one important cause for the dispersal of the chemicals. Figure 6.8 gives a rough illustration of how a chemical can be dispersed through the environment.

The **persistence** of a pesticide determines how long a chemical will remain in the environment or food chain. Pesticides are decomposed, rinsed away, evaporate, etc. When the chemical is broken down it usually becomes less toxic, but can be converted into a poisonous product (for example dieldrin, see appendix I). Decomposing can occur through a physical/chemical reaction (UV radiation, heating, reactions with other chemicals) or through biological processes (micro organisms). With vapourization the chemical evaporates from the soil and enters the atmosphere, the chemical is adsorbed to particles of the soil or it is incorporated into the food chain. All these processes (with the exception of efficient decomposition) simply shift the chemical and, therefore, also the problem.

In the end all pesticides decompose (with the exception of chemicals containing mercury, zinc or copper, etc.). The speed at which this break-down occurs, however, varies greatly. Natural pyrethroide compounds have a half life of some hours in sunlight and the active period of such compounds is one week maximum. The half life of certain organochlorines must be expressed in terms of months up to decades. Whether or not the rapid or slow break-down of a chemical is desired depends on how the chemical is applied: for instance, the protection of wood against injury from fungi or termites requires a slow break down. However, it is generally seen as a negative characteristic if a chemical can still be detected in a field after the end of a growing season.

Toxic effects on human beings

The risks for those working with toxic pesticides are great. Apart from those carrying out the spraying, the mixers and pilots of spraying airplanes are affected and there is substantial risk that others living and working near the treated area (farm-laborers working among the crops during or after spraying, villagers, and especially children) will unwittingly come into contact with the pesticide. Very little thought is given to these groups – especially where spraying from aircraft is concerned.

The risks are increased still further if the chemicals are inexpertly applied. Incorrect handling during application i.e. accidental switching of bottles, not observing safety use-by-dates, or accidents in the preparation or transportation of pesticides can have fatal consequences.

The effect of pesticides on health can be roughly divided into two categories: acute and chronic effects.

Acute effects can appear any time from a few minutes to a few days after contact with a chemical. They can vary from headaches, eye watering and skin blisters to more serious symptoms such as an breathlessness, dizziness, fainting spells, nervous tics or a breakdown in coordination.

The level of acute toxicity is expressed in: LD-50 (Lethal dose – 50%). This figure indicates what dosage of the chemical caused the death of 50% of the experimental animals, usually rats. These experiments generally take account of a period of between 2 and 10 days. The LD-50 is expressed in mg administered chemical per kg of body weight of the tested animal.

Figure 6.9. Two warning signs often used on labels of chemicals to express acute toxicity. The skull indicates a toxicity by which 0-3 grams of the chemical are lethal, an St. Andrew's cross indicates a fatal toxicity of 3-30 grams.

The **smaller** the LD-50 the **greater** the toxicity of the chemical!!

The route by which a poison enters the body is important, so the LD-50 is assessed separately for oral intake (through the mouth), dermal intake (through the skin) and intake through the respiratory openings. In general oral intake is the most dangerous. With chemicals easily dissolving in fat intake through the skin also involves major risks.

The toxicity of a pesticide differs somewhat from organism to organism. Parathion for example, is poisonous to people and other mammals, but not for plants. In appendix I for various organisms LD-50 of some pesticides can be found.

Chronic effects can only be observed after a longer period and often occur after long-term exposure to the chemical. Even if no further exposure takes place, the effects do not diminish, or do so only slowly. Chronic effects cannot usually be quantified as with the LD-50 in the case of acute effects.

The most prominent chronic effects are allergic reactions, damage of the generic material (mutagenity), damage to the fertility of both men and women, damage to an unborn child (teratogenity) and inducing cancer (carcinogenity). The symptoms often occur within the span of one lifetime although some of the symptoms are of consequence to progeny.

No doubt about one thing: the effects of pesticides are uncertain. What is regarded as being safe today can be found unsafe tomorrow on the basis of new research. One shortcoming of tests is that numerous chemicals may appear harmless while later they show to possess substantial carcinogenic characteristics. One example of this are the arsenical compounds.

6.2.3 Residue tolerance and safety periods

It is of great importance for the safety of both people and animals to know what happens after a pesticide has been used. We have already mentioned that some pesticides (DDT and allied chemicals) can accumulate in the fat of living creatures and for this reason, the use of this type of persistent chemical is forbidden in most western countries. There is currently a mandate on every pesticide requiring that it be decomposed within a certain time limit, into parts dangerous to neither people nor animals. This applies not only to that portion sprayed on the plant, but also to that on the soil.

On food crops the compound must be decomposed to such an extent that any residue is no longer dangerous for the consumer. The level of this residue is known as **residue-tolerance**.

Separate residue-tolerance levels have been established for the various chemicals and for the different agricultural and horticultural produce (see table **). The residue-tolerance is expressed in mg of chemical per kg of product (parts per million, ppm).

Practical application of the figures given for residue-tolerance is often difficult, so a so-called safety-period has been established for certain compounds. The safety-period is the interval in days or weeks which must elapse between the last treatment of a crop and its harvest. If a farmer holds to this safety-period, the residue-tolerance will not be exceeded. It must be noted that residue-tolerances are arbitrary: they vary greatly according to country and they are subject to rapid change.

Example

Chlordane is very poisonous for water-organisms and is also exceptionally persistent and therefore difficult to breakdown. In water Chlordane is initially absorbed by plankton but may subsequently be found in increasing concentrations in small fish, larger fish, birds or other fish-eaters and, finally, in man. Animals which are suspected to have died from poisoning should, therefore, never be eaten.

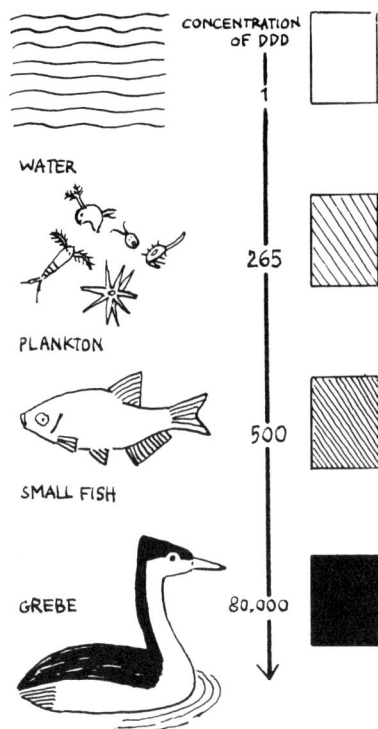

Figure 6.10. Accumulation of a pesticide in the food chain.

CONCENTRATION OF DDD

WATER — 1

PLANKTON — 265

SMALL FISH — 500

GREBE — 80.000

Table 6.1: Residue tolerance of some chemicals

Chemical	Product	Residue-tolerance (ppm)
MPCA	potatoes	0.1
	grain	0.05
Paraquat	potatoes	0.1
	grain	0.05
Parathion	potatoes	0.01
Lindane	potatoes	0.01
	grain	0.5
	flour	0.05

Another important characteristic is the harmfulness of a chemical to other organisms. This characteristic determines how much damage a residue in the environment will cause to flora and fauna (including cattle!). Such damage may manifest itself as acute toxicity, i.e. the death of fish caused by the presence of endosulfan in surface water. It may also become evident by accumulating in the food chain (figure 6.10).

It goes without saying that the "modern fishing technique" which involves putting a can filled with endosulfan in the water and returning the next day to scoop up the mass of dead fish is also very unwise.

The pollution of the often scarce waters of the developing countries is a particularly serious threat to the environment and in general the pollution of water should be scrupulously avoided. In situations where contact with water cannot be avoided, such as in the spraying of wet rice crops and the control of the malaria mosquito, it is vital that a chemical with a rapid break down period is used.

6.3 Safety measures

Structural measures such as adapting the law, forbidding dangerous chemicals, introducing cultivating practices which prevent pests, training concerning integrated pest management and supplying medical facilities in the event of accidents with pesticides, are all measures successful in the long-term.

Short-term effects can be obtained by beginning to use pesticides safely. For more extensive directions on safe use of pesticides see appendix I.

6.4 Formulation and form of a pesticide

The formulation of pesticides. Pesticides are never sold in their pure (active) form. Many other chemicals have usually been added which give the active chemical a particular structure and extra characteristics so that they can work better in practice. The whole process is known as "formulation". Possible additives are: detergents (substances to lower the surface tension), dissolvants (e.g. oil), emulsifiers, synergists etc.

The form of a pesticide. Pesticides can be offered in a variety of forms: liquid, powder or

Example

Echinogloa crusgalli (crow's leg) is, like maize, resistant to Atrazine. If Atrazine is used as a herbicide in maize and Crow's Leg is prevalent in the region, another compound must be sprayed to stop the effect of the Atrazine being totally negated.

63

Figure 6.11. Formulation of a pesticide. The active ingredient is mixed in a special way with other substances, like carriers, solvents, detergents, etc. This mixture usually is solved in water by the applicator.

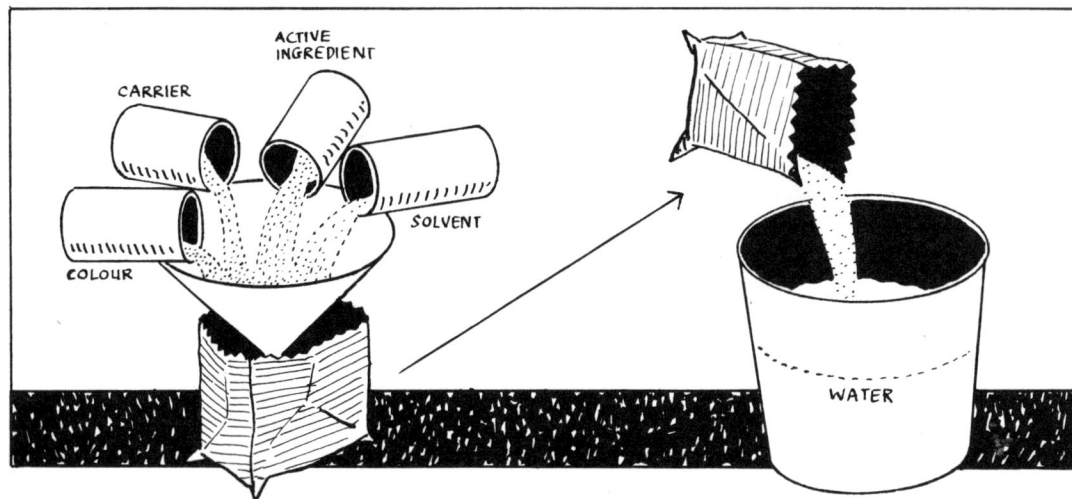

granules and a single pesticide is sometimes available only in one form, sometimes in several forms. If the pesticide is in liquid or powder form, it must be diluted usually in water, and sprayed. When applying pesticide, the liquid forms are preferable to powders. Although liquid forms are usually more poisonous, there is less chance of the chemical dispersing while preparing the spraying solution, than with powders.

Granules are suitable for applying to the soil. Those which are worked into the soil with the aid of special apparatus are even safer. They can, however, be impractical and have the disadvantage that while they can certainly be used in irrigation agriculture, it is afterwards impossible to walk barefoot on the treated fields. The use of granules also prevents the potential use of paddy rice fields for fish culture.

6.5 Possibilities for the efficient use of pesticides

There are many possibilities for minimalizing the use of pesticides and its related risks in crops. Pesticides should be used as efficiently as possible. The efficient use of pesticides means using an appropriate pesticide in the correct formulation and in the proper amounts, applying them only where it is necessary, at the proper time, in the proper place, under the proper weather condition. Applying a pesticide properly is therefore anything but simple. Below you will find a number of practical tips for this.

6.5.1 Identifying the pest

It is of the utmost importance that the pest is correctly diagnosed. Has the damage really been done by the suspected insect? Are the symptoms of the disease caused by fungi or is it a deficiency disease? If there is any doubt, an expert or a good manual should always be consulted (see appendix II). Only when there is absolute certainty about the cause of the damage chemical treatment is advisable, otherwise the treatment will be ineffective or may even have adverse effects.

6.5.2 Determining whether a pest exceeds the damage threshold

For an extensive description of the use of thresholds when deciding whether or not to apply chemical treatments see paragraph 6.1.

6.5.3 Selecting the pesticide

When choosing a pesticide for treatment the following points should be taken into account:
The proper pesticide for the pest to be eliminated. Insects should be treated with insecticides, weeds with herbicides, etc. This is almost entirely subject to the correct identification of the type of pest. There are more than enough examples of the wrong pesticide being applied, such as insecticides against fungal diseases or against deficiency related diseases. It may also be worth considering a selective pesticide (see 6.2.1) if available. Other properties of importance are: persistence of the pesticide (in most crops pesticides with a short half-life do well), toxicity for consumers, risk of development of resistance of the pest etc. Appendix I gives a list of fifty commonly used pesticides together with their target-organisms.

Properties of the ideal pesticide

An ideal pesticide should:
- control the target organism (i.e. a certain pest);
- not be toxic for the current nor for the next crop;
- not be poisonous to man;
- be selective (killing only the target organism and not affecting other organisms);
- be stable to protect the crop for a long time;
- break down fast into harmless substances in order not to contaminate food or the environment;
- not accumulate in food chains;
- keep its toxic effects to the target organism when applicated repeatedly (the target organism should not develop resistance);
- not be mobile in the environment.

Selecting a good form and formulation. The form and formulation (see 6.4) of a chemical generally has a lot of influence on the manner of application (spraying, working into the soil, etc.) and on the manner in which the pesticide comes into contact with the pest. A good pesticide dealer can provide further information.

Furthermore, it is also wise to purchase no more of the chemical than is required for one treatment, or than will be required for one growing season. Many pesticides cannot be preserved indefinitely and storage always involves risks.

Applying a minimal dose. In many cases the dose of pesticide sprayed far exceeds the amount necessary to eliminate the pest. A lower dosage is often enough to control one pest, while the effect on the natural enemy is minimal. The directions for use provided with the pesticide indicate a dosage in general that is guaranteed to work. Farmers, however, have a tendency to "add just a bit more" which rarely or never has a positive effect. Used efficiently, it is often possible to use less than the recommended dosage; although in such an event the amount should first be tested.

Using a minimum dose of a pesticide does bring the danger with it of under-dosing. The pest is not controlled sufficiently and the treatment may well contribute to the development of resistance against the pesticide.

Alternate use of different pesticides. To prevent the build-up of resistance to pesticides in a pest, it is recommended that pesticides with varying working-mechanisms are alternately used. Pests rarely develop resistance to two pesticides simultaneously: thus if one individual in a population is resistant to pesticide A it will be killed in the following spraying by pesticide B making it no longer necessary to increase the dosage of pesticide A.

Using well-adjusted spraying equipment. If spraying equipment is not properly adjusted or maintained, the pesticide is not efficiently applied and there is a chance that the crop will be damaged. Moreover, an unnecessary high level of pesticide is used.

Maintenance of spraying equipment

A relatively simple and efficient method to reduce the use of pesticides is buying good spraying equipment and maintain it well.

Main goals of maintenance are:
- to keep the system running in an optimal way so that a regular spray can be spread by proper use of the sprayer;
- to prevent the equipment from leaking, in order not to spoil pesticides and to prevent contact of the user to pesticides.

For this purpose:
- washers of pump and lance should be checked and replaced regularly;
- parts subject to corrosion should be repaired or replaced, especially of the pump;
- the nozzles should be checked and cleaned regularly, so that they will produce a fine spray and no drops fall out (which usually do damage to the crop).

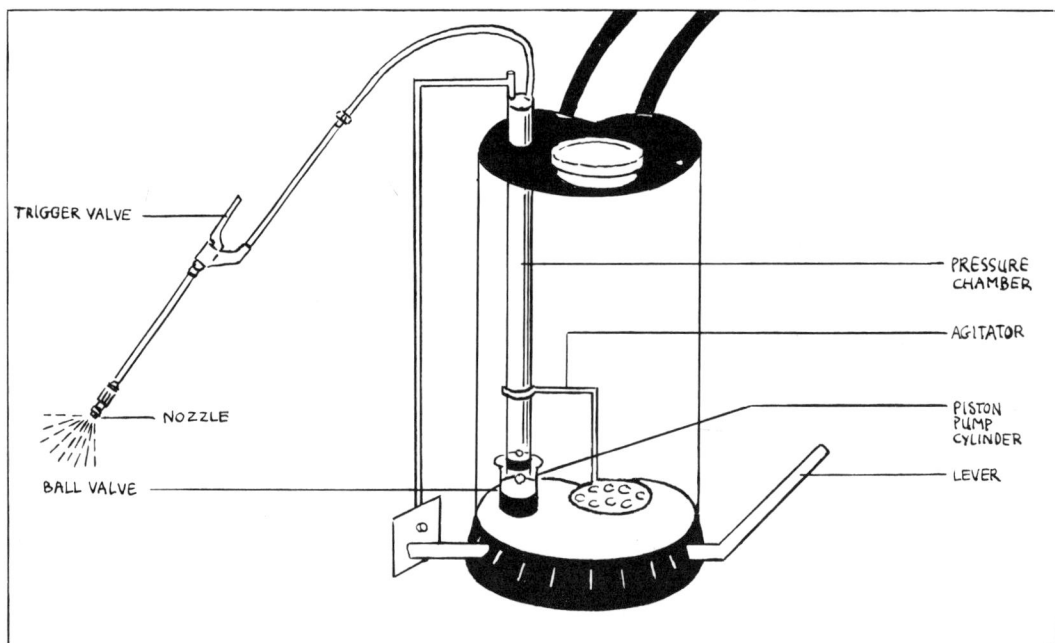

Figure 6.12. The knapsacksprayer and some of its main parts.

The knapsack sprayer

One of the most widely used pesticide applicators is the knapsack sprayer. For that reason some important features to be aware of when buying one are noted down here.

The tank.

- The tank should fit comfortably on the sprayer's back, by adjustable strings made of rot proof webbing wide enough (40-50 mm) not to cut the neck.
- Tanks out of metal are heavier, and more subject to corrosion than tanks out of plastic.
- The total weight of a full tank should not exceed 20 kg, so a content of 15 litres for the tank is OK.
- The volume of spray should be indicated by moulded figures on the tank.
- To facilitate filling, the tank should have a large filling opening of at least 95 mm in diameter at the top, also handy for cleaning the tank (with a gloved hand!!). The opening should have a tight fitting-lid to prevent spray liquid splashing out and down the operator's back.
- A filter (at least 50 mm deep into the tank) is needed to be put in the filling hole, to prevent debris (like leaves or small branches) blocking the pump or lance.
- When suspensions will be sprayed, an agitator in the tank is essential to reduce settling out.

Pump and pressure chamber.

- The size of the pressure chamber should be as large as possible, at least ten times its pumping capacity.
- The pressure chamber and pump should be fitted so, that maintenance and cleaning are easy to carry out.
- For lever-operated sprayers: under-arm levers in the long run are more comfortable than over-arm levers.
- Replacement parts must be available, especially of washers and parts suspect to corrosion by abrasive materials in pesticide formulations.

Lance.

- The lance should be fitted robustly and without leaking to the pump, the rubber connection should be easily replaceable in case of damage.
- The handle of a trigger valve on lances should fit comfortably and have a clip mechanism to hold the valve open for a prolonged time.
- Replacement parts must be available, especially of washers.
- When the spraying equipment are used for a certain goal (e.g. application in trees, or at the stem of rice plants, special lances usually are available to improve efficiency and reduce risks for getting in contact with the pesticides.
- Spare nozzles with varying spraying capacity should be bought together with the equipment.

Of some fifty sprayers tested in Central Africa, only four were sufficiently robust and recommended to farmers. The main faults were: poor linkage, inadequate strength of the metal tanks, the poor capacity of the pumps and poor strength of straps and their linkage to the tank.

Calibration of the sprayer.

The output of the sprayer can be controlled by measuring the spray liquid for 1 or 2 minutes. The pressure in the tank should be approximately like when in use. Having determined the output from the nozzle in litres per minute, the rate per unit area can be calculated, knowing swath width and walking speed.

With a swath of 0.8 m and walking at 50 m per minute and a flow rate of 0.6 litres per minutevolume of spraying per square meter is

$$\frac{0.6}{0.8 \times 50} = 0.015 \text{ litres/m}^2$$

which is (x10,000) 150 litres per hectare.

If the application rate is incorrect, other nozzles should be tried.

When the correct nozzle is selected, the volume applied can be rechecked by measuring the distance walked and the time taken to spray a known quantity of pesticide, again calculated towards litres sprayed per hectare.

Figure 6.13. If a pesticide is applied during the heat of the day, it may be decomposed by the sunshine, it may be phytotoxic and air movements make the chemical drift away. Therefore it is advisable to spray at the end of the day.

Figure 6.14. Spot-spraying reduces the use of a pesticide. Control of weeds in particular may be carried out in this way.

6.5.4 Proper timing of application

A pesticide should be applied at the moment it will have the greatest effect. The weather, the time of day and the season all play a part in this.

Spraying under the proper weather conditions. If spraying takes place when there is too much wind, the pesticide is easily blown away. If spraying is done in rainy conditions the pesticide may be rinsed away. Some pesticides easily decompose in sunlight and others show phytotoxic effects in that case. So spraying in the sun should be avoided and, moreover, in the sun it is far to warm to wear protective clothing.

Time of day. In the tropics it is advisable not to spray when the sun is at its hottest. The movement of the air at such times causes the droplets to rise and so prevents the greater part of the pesticide from reaching the crops. Spraying early in the morning, at the end of the afternoon or at the beginning of the evening is better.

Because of their activities, some insects are more easy to reach with an insecticide at certain times of the day than at others, and this should also be taken into account when deciding when to spray.

Example

The beetle **Gonecephalum simplex** *is found on coffee plants in Kenya. The adult beetles feed at night, therefore a mixture of insecticide and bait used to control this beetle are best applied at the end of the afternoon or at the beginning of the evening.*

Time during the season. Most pests are only found in a crop during one particular phase of the growing season making the timing of application of a pesticide very important. Many insects (e.g. stem and fruit borers) creep into the stem or fruit at the beginning of their larval stage, where they are difficult to reach. The larvae of moths often pupate in the soil and are therefore also inaccessible to pesticides. The most effective time for applying pesticides is at the most fragile stage of a pest's development i.e. after the egg has hatched and before larvae have bored into stem or fruit or pupate in the soil.

For the more short-lived organophosphorous pesticides it is important that the insecticide comes into contact with the pest while it is still potent.

Example

The **Anthores leuconotus** *beetle is found on coffee plants in Kenya and can be eradicated by smearing the bark of the coffee bush with a persistent insecticide before the rainy season. During this period the adult beetles appear and lay their eggs in the stem and therefore come into contact with the insecticide.*

When applying insecticides account should also be taken of the period during which the useful organisms are active. For example, in order to protect bees, spraying should not be undertaken during the flowering season.

6.5.5 Places of application

Pesticides should come into contact with the organisms they are to destroy. If the pesticides are only sprayed where the greater part of the pest is living, considerable savings in the use of the pesticide can often be achieved.

Spot-spraying. In spot-spraying only a part of the crop is treated with pesticide. This makes it

Figure 6.15. Some pests live on just a part of the plant. Application of the pesticide only on the infected part reduces the quantity of the pesticide being applied. In this case, a tail-boom adaption is used. If not available as such, a blacksmith may be able to make the construct one.

possible for instance for natural enemies to spread from the untreated areas and to re-establish themselves in the sprayed areas of the field. Used in conjunction with the elimination of weeds, spot-spraying can mean considerable savings in the use of herbicides.

Example
The beetle **Lissorhoptrus oryzophilus** *is found at the edges of rice fields, so if insecticide against this beetle is applied only to these edges, chemical can be saved.*

Many types of termites can be eradicated by applying insecticide solutions at the tunnels of the nests. The insecticide solution should be poured into the vertical nest shafts, preferably in the rainy season.

Example
The caterpillars of **Zeuzera coffeae** *make tunnels in the branches and stems of coffee plants. These caterpillars can be killed by introducing an insecticide into the tunnel before sealing it properly with something like clay.*

Place on the plant. The chemical should be placed on that part of the plant where it is most likely to come into contact with the pest organism:
- for leaf-eating insects the pesticide should either be applied to the leaf surface or in the leaf tissues;
- for diseases usually a systemic (a pesticide which is absorbed by and distributed over the plant tissue), fungicide or bactericide is required, these types of pesticides are usually expensive;
- systemic pesticides can also be useful against sap-sucking insects, in which case the pesticide should enter the phloem-system;
- for leaf-miners the poison should penetrate into the leaf-tissue in order to be effective;
- a soil-inhabiting pest can be attacked either through the tissues of the roots or else by contact pesticides introduced into the soil around the roots.

Trap crops. A crop more attractive to a pest organism than the cultivated crop, can be used to lure the pest away from the crop. A trap crop can be a variety of the crop itself or it can also be another host-plant. The pest then has to be killed before it has a chance to multiply to such levels that it spreads back to the crop. This process can be kept relatively simple and inexpensive either by using a small quantity of pesticide or by destroying the plants.

Trap plants are planted close to the fields (on dikes, alongside paths, on ridges built to prevent soil-erosion, etc.). The use of trap plants is expensive because they take up arable land and require investment of both time and money and by no means always produce a harvestable product. The introduction of a trap plant, however, is not nearly as radical or complicated as changing the crop-rotation or altering the crop-combination in mixed cropping.

Which trap plant could be considered; where and when they should be sown and treated or destroyed must be determined by experimentation carried out by expert investigators.

In many instances farmers already have an idea which plants attract certain insects. Moreover, more and more information on this subject is becoming available in the relevant literature.

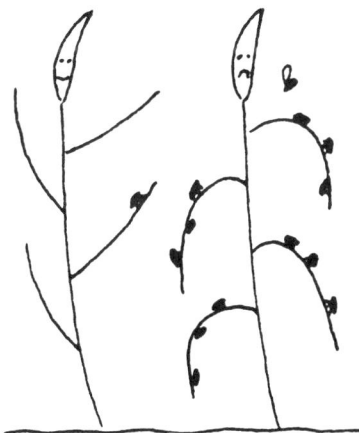

Figure 6.16. A more susceptible variety can be used as trap crop. Spraying of the latter kills a great part of the pest population.

69

Figure 6.17. Treating seedlings with a pesticide does not cost much and will prevent the field from being infected with soil pests. A pesticide to be used in this way should not be dangerous to man, for the planters will get in touch with it.

Example

The infestation of a rice field by **brown plant-hopper** *can be avoided by planting a variety of rice attractive to this pest organism around the edges. Because it is more attractive, the BPH will first establish itself on the susceptible variety which can then be treated with insecticide.*

Example

Patanga succincta is found in North-Thailand in maize fields amongst others. If a mixture of maize and soya are cultivated together, the insect will search for shelter in the soya bean in higher temperatures. It is then possible to destroy the insect by spraying the soya.

Treating seeds or seedlings. Through preventive treatment of seeds or by dipping the roots of seedlings in an insecticide solution, contact between useful organisms and the insecticide is also avoided. Treating seeds or seedlings with pesticides has the advantage of requiring small amounts of the relevant chemical. Moreover, the precautionary measures involved in carrying out seed treatments are less cumbersome than those for the spraying of the crop. When handling seedlings one should be aware of the dangers of phytotoxicity. The fragile and damaged tissue does not have a high tolerance level.

The treatment of seeds is often already carried out by the company which supplies the seed, and takes place under controlled conditions. This means less risk for the users. There are moreover simple instruments available with which seeds can be treated by the farmers themselves. For small amounts, jam jars in which the seeds and the pesticides can be shaken up are sufficient. This is relevant in those regions where farmers select and reserve their own seeds for sowing.

Example

Aleurocanthus woglumi is an aphid which is found on certain types of citrus plants and on coffee. Plant material can be protected by totally stripping it of all leaves and dipping it in an insecticide solution before planting.

Example

Platyedra gossypiella in cotton. The larvae of this butterfly start feeding at the end of June. The larvae frequently tie the petals of the developing buds together so that the flowers develop abnormally forming a rosetted bloom. Later in the season, the larvae burrow into the bulb and through the lint, searching out the seed. The feeding larvae destroy large quantities of the seed, the immature lint is weakened and stained and the young bulbs do not open. There are four or five generations per season.

P. gossypiella cannot be controlled successfully by the application of chemicals. Its main wintering quarter is the seed. The offspring and the extent of the infestations of this pest can be reduced by seed treatment. The seed should be put into water one day before treatment, and should be treated with heat – for 2.5 min. at 65 °C or for 1 min. at 72 °C. In addition, plant residues (where some 15% overwinters) should be destructed before the end of March. By the above described seed treatment, a moth called A. ypsilon attacking seedlings is controlled as well (see also paragraph 6.1).

7.   Storing the harvested products

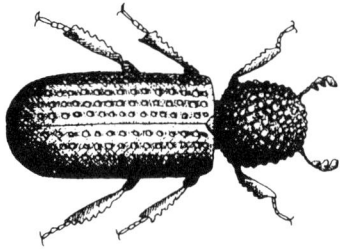

Figure 7.1. The lesser grain borer is one of the insects which threats stored grains.

In agricultural development policy a lot of attention is focussed on increasing production and scarcely any upon the significant contribution which could be made by protecting the harvest. Especially in cases where old varieties, which are often resistant against storage pests, are replaced by new, susceptible strains, the storage damage may be considerable. It is estimated that storage losses in the tropics amount to between 25 and 40% of the total harvest. Many of these losses can be prevented with a number of relatively simple measures, and therefore this chapter considers in some depth the correct storage procedures for products.

Losses in stored products are either caused by pest organisms or by physiological processes in the products themselves:
- attacking by insects or rodents;
- contamination by the excrement of insects or rats;
- ageing, sprouting or germination.

Agricultural products can be divided into **cereals and pulses** which can be stored for lengthy periods without special treatment, and **tubers, vegetables and fruit** which quickly rot. Just as the major causes of losses are unique to each of these groups, their respective storage methods also differ and are described in 7.3 and 7.4, while the storage of seed is discussed in 7.5.

7.1 Some general rules for limiting storage losses

On the whole, many of the losses incurred in stored harvest products can be prevented by attending to the following points:
- hygiene;
- correct storage climate;
- protection against rats.

Hygiene. As harvested products are usually infected while still in the fields, the first step towards preventing storage losses is to control pests during the growing season and to practice hygiene when handling the harvested product. Taking hygienic measures in and around the storage facility also prevents storage losses. Storage facilities must be thoroughly cleaned before any product is placed in them, and it is advisable to burn the collected rubbish in order to destroy possible germs and insects. Any holes or cracks which could provide shelter for insects should be filled in. Painting the storage facility white will show up dirt on the inside and on the outside it will keep the facility cool.

Finally, it is essential that the new harvest is **never** stored in facilities which still contain the remains of previous harvests. Pests which may originally have been few in numbers, could have multiplied in these remains and these could then form a source of infection for any new harvest stored there.

71

Figure 7.2. Grain can be stored in all kinds of sheds. This kind is protected against rats by collars, weeds around the shed are cut so that rats have no chance to hide from dogs or cats.

Climate in the storage facility. Relative humidity and temperature in the store, and the humidity of the product to be stored determine to a large extent which pests are able to develop. For example, fungi and bacteria breed well in warm and humid conditions. The optimum moisture level of the product to be stored varies greatly according to type (see 7.3-7.5).

Protective measures against rodents Various methods have been developed to limit losses caused by rodents. To keep out rats, storage facilities are built on stilts fitted with caps (fig.7.2), or products are stored in cans or containers, behind mesh or in bamboo boxes.

The immediate vicinity of the storage facility should be cleared of all weeds to deprive rats and mice of possible shelter, in combination with keeping dogs or cats this method can be very effective. If there is a serious threat of a plague of rats, overhanging branches can be removed. This must be weighed against the advantage of the shadow provided by these branches. As a last resort, traps can be set or poisoned bait may be used.

7.2 Applying pesticides in storage

If other methods do not suffice, they can be supplemented by chemical treatment of the storage facility or the product. Great care should be taken when applying pesticides to a stored harvest: there are always risks that something goes wrong during application, or that the product is not properly washed before it is consumed. Several methods of applying pesticides exist:

Disinfecting storage facilities. Disinfectants can be sprayed on walls, floors and ceilings. Malathion, tetrachlorvinphos, pirimiphos methyl, chlorpyriphos methyl (2-4%) and pyrethrum-piperonyl butoxide (0.5%) can be applied. Surfaces of grain-piles or heaps of bagged cereals inside the warehouse can be treated with 2% malathion to prevent contamination from outside. Insects already established in the grains are not killed by surface treatments.

If infected sacks are used, they can be treated with 2-4% of malathion, pirimiphos methyl or tetrachlorvinphos and must be air-dried after treatment. Cereals stored in 2% treated sacks can be stored for 3-5 months, in 4% treated sacks 6-12 months, depending on the type of cereal and insect infection.

An alternative to chemical treatment is washing the sacks in hot water (75 °C or more) and drying them thoroughly afterwards.

Fumigation. Fumigants are chemicals which, at the required temperature and pressure, can exist in gaseous state. They are able to diffuse and penetrate tightly packed materials to kill off a given insect. After the gas is dispersed, some hours to some days after application depending on the silo, the product is susceptible once again and new infestation must be prevented. Fumigants usually do not leave any residues. Pests which are introduced from the field may be effectively treated by fumigation. Fumigants are available in solid, liquid and gas formulations.

Most fumigants are inflammable vapours highly toxic to humans and animals, therefore they should usually only be applied by skilled operators, and are not suitable for use in small-scale agriculture. Only phosphorous hydrogen (a broad spectrum insecticide) which is available in small pellets is so simple to apply that as long as air-tight silos are in use it can be used by unskilled farmers.

Some storage pests have succeeded in building up resistance to fumigants, as is also the case for many other chemicals.

Protectants. A protectant is any chemical which is mixed with newly harvested products to

72

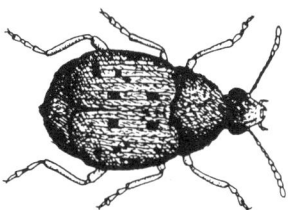

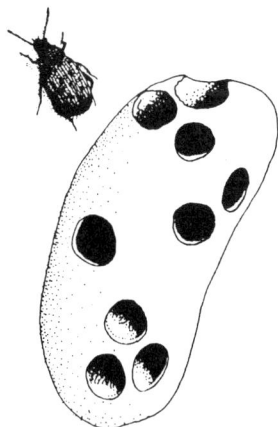

Figure 7.3. The bean bruchid, a pest attacking stored and dry beans.

prevent attack and infestation by pests. At present, the only approved protectants for cereals for human consumption are limited to a mixture of pyrethrins and piperonyl butoxide (0.05%) and malathion at 10 ppm active ingredient (1 mg/kg cereals).

7.3 Cereals and pulses

If cereal and pulses are well dried they may be stored for up to a year. Dry storage impedes bacterial infections and fungal growth. Fungi especially can produce poisonous substances (i.e. carcinogenic aflatoxins in peanuts) and cause serious damage to product quality.
Every product has a moisture percentage below which no rotting takes place. Products should be dried to this level.

Some of the 'safe moisture percentage' values are:

Product	safe moisture percentage
Cereals	12-14%
pulses	13-15%
Seeds containing oil	6- 8%

To bring the moisture below the 'safe moisture percentage' artificial drying is sometimes required.
Temperature fluctuations within a parcel of cereals can cause the air in the warmer places to absorb more moisture. This warmer, humid air rises to cooler places where the moisture again condensates. Such areas are susceptible to rotting and contents of the whole silo can be lost. By ensuring that the product and silo are properly dried and a relatively constant temperature this kind of loss can be prevented.
Cereals and pulses are usually dried in the fields, sometimes laid out on racks or bundled in sheaves. After threshing, the harvest is dried once more, either artificially or on canvas in the sun. Early harvesting necessitates artificial drying.
The various drying methods work by means of air-streams either created by wind or by ventilators, these must be able to blow through the whole parcel, otherwise the product will not be properly dried. The best method is to spread the product out in equal layers that are not too thick, at right angles to the wind. Special platforms or stages can also be used.
Once dried, the product must be allowed to cool off before being put in the container, otherwise the above described process may cause rotting of the lot.

7.3.1 Storage methods for pulses and cereals
The methods for storing cereals and pulses vary according to whether the product has been threshed or not.
Storing unthreshed cereals is carried out in what are known as maize cribs (see fig. 7.4). The width is determined by the climatic conditions. The maize dries fairly quickly from a moisture percentage of 35% to 14%. The product can also be stored in baskets as long as the atmosphere is not too humid during the storage period, the baskets themselves frequently being stored in huts.
The above mentioned methods offer little protection against insects and rats, although rat-

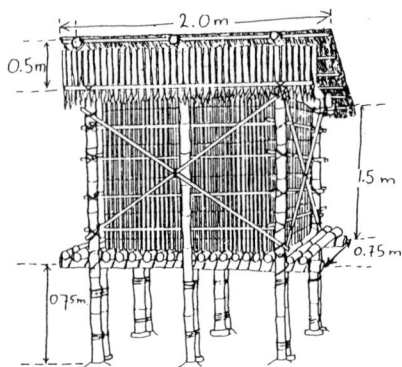

Figure 7.4. A crib made of bamboo, suitable for maize and yam.

collars placed around the stilts of the huts and cribs do help to a degree. If whole, the husks and chaff of maize, rice and beans do provide some natural protection. Many modern varieties however no longer have a sealed husk, and there is also the danger that if the moisture percentage is too high, fungi will grow under the husk.

If the husk offers no protection then application of a pesticide can be considered. For a pesticide to work effectively the husk or pod must be removed to allow contact to the grains.

Threshed products can be stored in several ways depending on the relative humidity of the atmosphere during storage. During humid storage periods cereals and pulses should be stored in water-tight containers. According to the quantity, these can be plastic sacks or metal or earthen silos. It is also important whether the product is to be used for consumption during storage. If so, the most commonly used silos are those fitted with a re-sealable vent or lid. If large quantities are to be stored untouched for long periods, then underground silos are frequently used.

In dry storage periods, and if properly stored, the product can dry still further during the actual storage-period. Suitable storage means for this purpose might be: baskets, earthenware pots or gunny sacks.

7.4 Root and tuber crops, fruit and vegetables

Root crops such as cassava, potato, yam and batata as well as fruit and vegetables require a quite different storage approach. These products contain a lot of water (60-80%). Drying is bad for the product quality, while too high a percentage of water causes rotting.

Root products also frequently have a high respiratory rate which results in a further rise in temperature and in the risks associated with this. Root products have a period of inactivity before sprouting. For yams this period lasts for 4 months while potatoes start sprouting after only 5 weeks.

When cultivating root and tuber crops it is particularly important to choose a variety resistant to storage diseases and pests. Choosing the proper harvest time is important: over-ripe fruit is difficult to store if unprocessed. Root and tuber products also have an optimum harvest time. Proper hygienic practices should be followed; damaged and infected fruit and tubers must be removed from the lot before storage and thereafter the stock must be controlled for signs of further infection. Sometimes the product is treated with pesticides against fungal infection or against sprouting with a growth-regulator.

Other methods for safe storage are:

Harvesting carefully. Greater care is required in harvesting a root-crop than with other crops because rotting always begins around wounds caused during harvest or transport. Damage can be avoided by using wooden tools devoid of sharp edges and points. Damaged root products are best consumed immediately unless they can be treated. Rubbing ash, loam or chewed cola-nuts into the wounds and leaving them to dry in the sun will reduce the chances of rotting considerably.

Curing. Root products are frequently subjected to a form of protection known as 'curing'. The products are stored for some days in warm and humid conditions (25-35 °C, 80-95% humidity)

74

so that a layer of cork-cells is formed which protect the product against excessive dehydration and infections. For the best result, this treatment should be carried out in shaded conditions to avoid too rapid a reduction in the relative humidity of the surrounding atmosphere. Some values for efficient curing are:

Product	temperature (°C)	rel. humidity (%)	duration (days)
Cassava	30-35	80-95	4-7
Yam	29-32	90-95	4
Batata	30	85-90	5-7
Potato	8-20	90	5-8

Applying suitable storage methods. The storage methods are especially directed to keep temperatures as low as possible. The lower the temperature the less chance of dehydration, rotting or sprouting. Although this can be difficult in the tropics, reasonable results can be obtained by using a ventilator. Ventilation takes place during the cool night hours, while during the heat of the day everything is sealed off so that the lower temperatures are retained. This does require storage facilities with thick, well-insulated walls.

Some rootcrops such as yam, batata and cassava can be left in the soil and harvested when needed. The disadvantages of this are that the ground is left unused for longer, the product can be damaged or stolen by rats, termites and apes and that the products in time become more fibrous. A storage method used for many roots and tubers is the 'clamp'(see fig. 7.4).

A number of storage methods are used for each crop, for example, cassava is frequently packed in cases in layers alternating with moist sawdust or cassava leaves. Using this method the maximum storage is 1-2 months.

In dry periods, yam is often stored in so-called yam-sheds, four-sided constructions made of latticed fencing with a palm-leaf roof; the yams being hung from the lattice work. This provides sufficient ventilation and the yams can be properly controlled. With adequate control and the removal of sprouting and rotting tubers the yams can be kept for between 3 and 4 months. Batata and potatoes are frequently stored in clamps and protected against rats by a covering of netting or bamboo. With this method it is important to ensure that ground-water and rain cannot reach the bottom of the clamp.

Processing. There are many methods for processing fruit, root- and tuber crops. Cassava, batata and potatoes, for instance, can be sliced, cooked and properly dried into a sort of crisp which, if packed in air-tight containers, can be kept for a long time. Produce can also be processed to jams and pickles.

7.5 Storage of seeds

The storage of seeds basically is like the storage of cereals and pulses, but puts extra demands on the available facilities. Hygiene must be more stringently practiced than when storing products for consumption, otherwise pests may be introduced into the field as early as the sowing stage. Retaining the germinal force often requires a precise moisture percentages. A rule of thumb when storing seeds is that the period in which the germinal force remains stable

Figure 7.5. Small amounts of grain, like seeds, can be stored in tins, earthen pots or bamboo containers, depending on materials locally available.

75

Bio-control of pests in stored products

There is an increasing need for alternatives to controlling insects chemically in storage because insects are now showing resistance to the most common storage pesticides – especially to fumigants such as phosphorous hydrogen. Furthermore, the technical possibilities for developing new pesticides here are much more limited than with insecticides intended for other situations, because the products are to be consumed readily. Natural enemies are still not being used to control storage infestations. This is rather remarkable as it concerns pests in simple, manipulatable and more or less closed-off ecosystems as in greenhouses where successes are notable. It must also be possible to use biological control to eliminate storage pests in storage facilities. Far too little research is being conducted in this area and therefore virtually no practical techniques have been developed.

doubles with each 1 percent reduction in the seed's moisture percentage within the range of 5-14% and again for each 5 °C reduction in temperature down to 5 °C.

Damage to seed can be caused by machines during harvesting or threshing, or by insects or rats. This damage does not only have a direct influence on the germinal force of the seed but leaves them more susceptible to fungi and bacteria.

In conditions of humidity above 20% the seeds will sprout. The safe humidity level for most seeds is between 7 and 9%, with the exception of coffee, cocoa, oil-palm, and citrus seeds which require a higher humidity level for storage.

When drying seeds, temperatures above 35 °C will reduce the germinal force. If the harvest-time is dry, it is recommended that the seeds are dried in a shaded place in the open-air. In wet periods artificial drying is unavoidable and should be done with great care, for instance with the aid of lamps, moderate ovens or water-absorbing materials such as clay or ash.

Applying pesticides to stored seeds is not quite such a drastic measure as with stored produce intended for consumption. If well selected and correctly applied, pesticides can control or eliminate many pests without usually having any influence on the germinal force of the seeds. Some pesticides do negatively influence the germinal force, i.e. Lindane, as do the fumigants (pesticides in gas form) methylbromide and ethylenedibromide (see frame).

Seeds are best stored in air-tight containers. Water-absorbing materials and a suitable pesticide such as malathion can be mixed with the seed. Common storage methods are strong plastic sacks, painted earthenware pots, varnish or linseed oil tins and steel containers with a screw-lid or bamboo containers.

Seeds are also frequently mixed with ash or dried clay to a factor of 1:1 and topped with a layer of ash two centimeters deep. This will keep the seeds dry and pests find it very difficult to penetrate such a mixture.

Pesticides for use in storage

There is a number of insecticides available for use during storage: malathion, pirimiphos-methyl, tetrachlorvinphos, pyrethrum/piperonyl butoxide as well as a number of fumigants. The most commonly used are: malathion in powder form and pirimiphos-methyl in powder or liquid form. See the pesticide list appendix I.

For for seed purposes only the following insecticides may be used: tetrachlorvinphos, pirimiphos-methyl, carbofuran and chlorpyriphos-methyl. These are effective against the common pests for 3 to 9 months of storage (the higher the dosage, the longer the effectiveness) but cannot be recommended for treating cereals for human or any consumption.

8. Conclusions

The previous chapters described many of the measures which could be taken to control pests. In this chapter we list these methods one by one, categorized according to the different type of pest.

8.1 Integrated weed control

Sanitation:
• use uncontaminated seeds and plant material.

Cultivation practices:
• sow the crop densely so that the canopy will close quickly.

Biological control:
• cultivate crops which quickly close their canopy;
• keep the soil covered by intercropping;
• grow a green fertilizer after the main crop;
• graze fields lying fallow.

Physical control:
• plough under weeds and prepare seedbeds immediately before planting or sowing;
• apply a mulch;
• hoe, harrow and earth up.

Chemical control:
• spray infected spots only;
• do not use any herbicide which is dangerous or persistent.

8.2 Integrated insect control

Sanitation:
• burn off old stubble;
• use uncontaminated seed (disinfect either by thermal treatment or chemically).

Cultivation practices:
• keep the nitrogen fertilization just below the optimum yield level (against aphids and spider mites);
• encourage natural enemies by using organic manure.

Biological control:
- cultivate resistant varieties;
- introduce natural enemies.

Chemical control:
- check crops regularly and spray only when damage thresholds are exceeded;
- consider well how to applicate insecticides: use only at the proper time and the proper place (sparing natural enemies); use the proper chemical in the proper dosage, only if there is an inadequate presence of natural enemies in the crop.
- prevent the development of resistance by alternately using insecticides with different working mechanisms.

8.3 Integrated fungi and bacteria control

Sanitation:
- use uncontaminated seed (disinfect with a thermal treatment or chemically).

Biological control:
- cultivate resistant varieties;

Cultivation practices:
- keep the nitrogen fertilization just below the optimum yield level;
- take care the climate inside the crop is not too humid by making sure the crop is not too dense.

Chemical control:
- check crops regularly and use damage thresholds;
- plan the application of fungicides: use only at the proper time and the proper place (sparing natural enemies); use the proper chemical in the proper dosage;
- prevent the development of resistance by alternately using fungicides with different working mechanisms.

8.4 Integrated soil-disease (fungi, nematodes) control:

Sanitation:
- use uncontaminated plants and seeds;
- clean tools and footwear before moving from one field to another;
- use uncontaminated compost and manure.

Cultivation practices:
- practice a healthy crop rotation scheme.

Biological control:
- stimulate an antagonism in the soil by using organic manure and green fertilizers;
- cultivate resistant varieties.

Chemical control:
• apply pesticides locally (e.g. in seedbeds) and use only admitted soil-disinfectants.

8.5 Integrated control of rodents

Cultivation practices:
• cut weeds in the vicinity of fields to destroy potential shelter for rodents;
• reduce the size of dikes in rice fields to cut down the space where rodents can burrow;
• plant rice simulataneously over great areas.

Biological control:
• keep trained domestic animals such as cats, dogs or non-poisonous snakes.

Physical control:
• dig out rat-burrows and destroy the inhabitants;
• use mungo between the crops to act as a barrier.

Chemical control:
• with the cooperation of other farmers, carry out sustained baiting from the start of the planting season.

Part III
Research, extension and small-scale farmers

This part discusses the position of integrated pest management in agricultural research, in extension services and in agricultural practices. It first considers what is taking place at international level and then at national level in this field, concluding with a detailed description of how IPM fits in at farm level. The most crucial issue is after all to bring IPM into practice in small-scale agriculture.

9. Introduction to agricultural research and extension

Figure 9.1ab. Both farmers and scientists do experiments. Farmers in an informal way (adapting new ideas to their specific situation), scientists in a formal way (via standardized methods).

The introduction of integrated pest management will be more successful when the IPM methods are better adapted to the current situation, customs and practices of the small-scale farmer. An IPM programme must therefore locally be developed, not only because of the differences in climate and crop growth, prevailing pests and the presence of natural enemies, but also because of the technological level of the farmers and the existing customs and cultural practices. Applied crop protection research must therefore use the current situation as the starting point in order to achieve serviceable results.

The history of crop protection technology, however, is marked by the export of successful control methods from the West to the Third World. This implies that Western methods are made applicable at a research institute in the Third World, and the extension services must then present these methods to the farmer. Up to now the Third World has been confronted by this form of "development".

Research has proved to be essential for the development of new techniques in agricultural production, which has resulted in recent increase in production. For many small-scale farmers, however, agricultural developments have brought much that is new, but little that is good. The results are usually not only an increase in production, but an also increase in dependence on inputs and manufacturers, a disruption in social relationship within the farming community, and environmental pollution and health hazards. The realization is now beginning to dawn on the researchers, that for the above reasons, the farmer must be seen as the starting point of research, and not as the coping stone.

Research is, in fact, nothing more than the search for the solution to a problem. It is a normal human activity for anyone who considers a situation and tries to improve it, regardless whether the researcher is a farmer, an agricultural scientist or an extension officer.

However, a distinction can be made between formal and informal research. Formal research, as conducted at universities and (scientific) institutes, is directed towards standardization of methods and testing of hypotheses. Informal research, on the other hand, tries to find an appropriate solution for a specific situation or problem. Farmers are basically experimenters themselves, because their situation always has been subject to changes (political, economic, demographic, physical), and farmers must react to this.

It is very important for scientists and development workers to realize that research is not limited to only the process of "searching". Another part, which has just as vital an effect on the final result, is the process of making decisions and choices. Sometimes this is a conscious, but more often an unconscious process.

Essential questions in this respect are:
- What exactly is the goal of this research?
- At which group of people is it aimed?
- Is the problem seen as being entirely technical or also with social aspects?
- Is involvement of the target group desirable in determining the problem and in further research?
- How, and to what degree are the people to be involved?
- What view is to be taken of responsibilities towards the people for whom the research is intended? And towards the organization for which the work is carried out?
- What are the aims of this organization?

From the second half of the seventies onwards, many voices are heard demanding that farmers should be put into the centre of agricultural research, and thus also in crop protection research, which is linked to an important role for extension services. An intensive feed-back between researcher and farmer is necessary to achieve the successful introduction of integrated pest management methods to small-scale farmers.

Developments in agricultural research and extension on international and national level will be discussed in chapter 10 and 11. Chapter 12 deals with extension aimed at small-scale farmers. Part III concludes with indicating bottlenecks which can arise from these new approaches.

10. Research and extension at international level

Agricultural research in the tropics was originally carried out in colonial multidisciplinary research stations and was usually geared to high value export crops. These stations usually had no ties with the rural food producers, except when this was integrated with the production of the relevant export crops. After the success of increasing agricultural production in the United States a willingness arose to pass the new production technologies on to Third World. The knowledge acquired had to be transferred to farmers in Third World countries. In this period the model of the US Land Grant University became popular and had considerable success in the United States: "... (the model) combines education, research and extension; research to be guided by the practical experience of farmers." The United States was in this sense an important step ahead of European countries, which still had an unshaken belief in agricultural research as conducted at universities and institutes.

During the fifties it became clear, however, that even this new approach did not meet expectations in the Third World. The necessary services (government or private) were, and still are, not available to the majority of the farmers, and farmers also lacked significant political influence. Furthermore, the starting point of this approach was an increased efficiency of market oriented farmers, whereas less efficient farmers would disappear and provide labour outside the agrarian sector. It became clear that it was necessary to adjust the new techniques to local circumstances, especially to those of small-scale farmers: local knowledge and institutes expanded or were set up.

10.1 International Agricultural Organizations

10.1.1 International Agricultural Research Centres.

In the fifties the agricultural problems in the Third World, such as stagnating food production and increasing population pressure, were recognized. The inadequate provision of information was seen as an important cause of this. In this view, the agricultural research in the Third World needed to be boosted. This was the reason why a number of Western countries set up International Agricultural Research Centers (IARCs) with the primary aim to undertake fundamental agricultural research and to support local research.

In 1971 the CGIAR (Consultative Group on International Agricultural Research) was set up to coordinate the activities of what had by then become 13 IARCs. In 1987 these 13 IARCs had 243 million US dollars available, which had come from various donor countries, foundations (Rockefeller and Ford) and international organizations (such as the World Bank and UNDP). Each of these IARCs was given a specific task in terms of a crop or an agro-climatic zone (see table 10.1) and the research stations were aimed, and still do aim, to stimulate food production

WHEN BREEDING RESISTANT VARIETIES, ONE CAN USE EITHER PHYSICAL OR CHEMICAL RESISTANCE CHARACTERISTICS OF PLANTS

Figure 10.1. The problems tackled at international research centres usually are of a fundamental character.

in the tropics. These developments disassociated themselves from the research done at colonial research institutions where attention was mainly directed at export crops. The line of approach was also different from the US Land Grant University model, which exported ideas and methodologies from the United States to the Third World. The IARCs produced knowledge for the Third World and transferred this knowledge through communication media and by providing courses for local staff.

Example

The first IARC was the International Rice Research Institute (IRRI). IRRI was set up in 1960 in the Philippines on the initiative of the Rockefeller and Ford Foundations. Since its establishment, IRRI has grown to an organization with more than 2500 employees and in 1986 an annual budget of nearly 30 million US dollars. In 1966, IRRI produced the short-stem rice variety IR8, and by 1974 a quarter of the rice area in the Third World had been planted with this variety or one of its successors; the so-called high yielding varieties (HYV). Yields rose phenomenally where the artificial fertilizers and pesticides required by HYV were available. This development was called the "green revolution".

The green revolution also had its negative effects. Rice production did indeed increase, but social contrasts increased in rural areas leading to an exodus of people who could not raise enough capital. These problems persist up to now. This criticism has led, through the years, to a change in direction and particularly to an expansion of IRRI's research towards devoting more attention to the social aspects of rice cultivation.

A general characteristic of the IARC's is that they are autonomous research and training institutes with their own management. Through their international status they are safeguarded against all kinds of national restrictions and bureaucratic procedures. They enjoy free exchange of information and within reason are politically independent. Current research is directed at one or more (usually food-)crops, or at certain climatic zones. Their financing is guaranteed by international funding agencies, so the institutes have good quality material available and can afford good (and expensive) staff.

In general there is a close link between the international institute and its host country. IARCs collaborate with local faculties of agriculture and assist in carrying out national programmes. Many IARCs have so-called "outreach programmes": regional research groups which transmit results through national organizations, help with the compilation of "recommended packages", to facilitate the application of the technologies developed and to stimulate and support regional programmes. For the IARCs themselves these activities are also of great importance because the technologies developed can be tested under different conditions, and feed-back is guaranteed.

Important production limiting factors (including pests) have a high priority with the IARCs. Research is aimed at:

- The development of resistant varieties: new varieties are developed, that in each country must be further adapted to meet local requirements. IARCs also maintain gene-banks to ensure that local varieties are not lost.
- Research into crop protection strategies. IPM (e.g. for rice at IRRI), and research into cropping patterns, intercropping, etc. are now the spearhead.
- The development of new technologies and concepts, such as screening techniques for the identification of certain diseases, and techniques to evaluate the resistance of new varieties.

Table 10.1 Some of the 13 International Agricultural Research Centers, their research objectives and regional focus.

IARC	Objectives	Regional focus
Centro Internacional de Agricultura Tropical (CIAT, Columbia)	Cassava, pastures, field bean	Latin America
Centro Internacional de Mejoramiento de Maiz y Trigo (CIMMYT, Mexico)	Barley, maize, triticale, wheat	Latin America
Centro Internacional de la Papa (CIP, Peru)	Potato, sweet potato	Third World countries
International Centre for Agricultural Research in the Dry Areas (ICARDA, Syria)	Barley, chickpea, faba bean, lentil, wheat	Third World countries
International Crops Research Institute for the Semi-Arid Tropics (ICRISAT, India)	Chick pea, ground-nut, millet, pigeon pea, sorghum	Third World countries
International Institute of Tropical Agriculture (IITA, Nigeria)	Cassava, cocoyam, cowpea, maize, rice, soy bean, sweet potato, yam	Sub-Saharan countries
International Rice Research Institute (IRRI, Philippines)	Rice	Third World countries
West Africa Rice Development Association (WARDA, Cote d'Ivoire)	Rice	West Africa

In addition to fundamental research, the IARCs also aim to develop techniques or methods which are broadly applicable. Adaptation to local circumstances and breeding is left to local research institutes and the actual transfer of new techniques to farmers is also outside their mandate. However, sociologists inside the IARCs are nowadays also working on this problem. They investigate why small-scale farmers apply the new technologies or fail to do so, and aim to link the research of the IARCs to the everyday practices of the farmer. They are thus trying to form a link between the small-scale farmers and the institutions which deal directly with them.

Figure 10.2. International research institutes are politically independent, are safeguarded against all kinds of bureaucratic procedures. They are financially dependent on stable donors so there is no lack of facilities and funds for expensive staff.

Another important function of the IARCs is the transfer of information. Thus, IARCs are concerned with:

- Research training for researchers working for national institutes. This is important to transfer new developments in techniques and varieties. Training is often organized in conjunction with universities, so a degree can be obtained after the training.
- Training in cultivation techniques for extension officers.
- Provision of library facilities, organization of conferences and other sources of information.

Developing IPM methods for small-scale farms is not a pressing item for the IARCs: they hardly are concerned with such small-scale affairs. They see their task mainly as one of support and provision of practical aid to national research institutes and extension services in the field of technologies for increasing production. It is obvious that the IARCs use a technological strategy to improve agriculture. Such an approach has its possibilities, but also its limitations, especially in case small-scale agriculture in the Third World is concerned.

10.1.2 Food and agriculture organization
The Food and Agriculture Organization (FAO) of the United Nations was set up in 1945 with the aim of improving food security in the world. Crop protection has always been of central interest to the FAO from the point of view that improving the health of a plant is very effective way to increase agricultural production. The FAO thus also considers it an important task to convince the authorities in the Third World of the importance of crop protection. The problems concerning pesticides also receive considerable attention from the FAO. Important activities in this respect are the registration and control of pesticides, the framing of the International Code of Conduct on the Distribution and Use of Pesticides in collaboration with other international organizations (see Part IV), and establishing standards of quality for pesticides and residual tolerances in collaboration with the World Health Organization.
In addition to this, the FAO has specific crop protection programmes. The best known is undoubtedly the FAO Desert Locust Programme. Usually this is carried out through national governments: FAO supports research (as well with staff as with materials) and extension (i.e. assistance with the organization of special extension campaigns, the supply of inputs).
The FAO also has some programmes to promote IPM, i.e. in the cultivation of rice in various South and South-East Asian countries.
Acting under the UN-flag, FAO has a lot of influence. For example, as a result of the efforts of FAO, IPM has been adopted as the official policy in Indonesia and President Suharto in 1986 prohibited the use of organophosphorous compounds (insecticides) in order to protect the natural enemies of pests in rice.

10.1.3 Agro-chemical industry
Multinationals such as Shell, Bayer, ICI, Ciba-Geigy and Dow Chemical are important producers of pesticides. The Agro-Chemical Industry (ACI) naturally directs its research mainly towards the development of pesticides and their application, as production of pesticides is very lucrative. The research of the ACI is mainly geared to Western agriculture where there is a constant demand for new products. Nowadays the demand for more selective, less persistent and less toxic products is constantly increasing. For the ACI, however, the development of broad-spectrum products is much more yielding on account of their wider field of application. The

*The FAO project "intercountry pro-
gramme for the development and application of integrated pest control
in rice growing in South and South-
East Asia" provides support to projects for integrated pest
management in rice (see also chapter III-4). This support concerns primarily the creation of an
infrastructure for communication
within and between the countries,
and between high level administrative staff and the extension services.
The exchange of information between farmers and extension officers is also an important aspect of
the programme. Many activities are
conducted at village level.*

ACI therefore concentrates on products which are not very persistent and toxic and to a lesser extent on selective products.

Research and development are very important for the ACI as a new pesticide can totally remove another from the market. A new active ingredient can be patented and monopolized for about 20 years. There is, however, a growing group of producers in Asia and Latin America who deny patents.

In addition to this, biotechnological research which is becoming more important is almost completely monopolized by the financially strong multinationals. The combination of interest in plant breeding and plant protection has led, for example, to the introduction of herbicide resistance in crops which allows an increased use of pesticides.

The ACI considers the Third World as a market with much growth potential and would like to reach this market by collaboration with the national governments. A pesticide producer can set up a network of commercial agents who will ensure that his products are brought to the attention of the farmer. Such a strategy is carried out in the Philippines by Bayer, where farmers have accepted pesticides on a large scale. In addition, the government provided free treatments for 10 years. The farmers often meet representatives of the chemical industry who use false reasoning to stress the need for the regular application of pesticides: they are sold under the guise of "yield booster" without explaining how the yield will be boosted. The situation in the Philippines is not exceptional: in many Asian countries, the governments have stimulated (and some are still stimulating) the use of pesticides, for instance by providing subsidies.

Small-scale farmers provide an interesting market for the ACI, like when trading pesticides which are forbidden in the West. In their research and development however, the ACI pays no attention to the problems of the small-scale farmer. The combination of the large-scale production of pesticides and the acquisition of seed breeding companies leads one to suspect the worst: the total control of agriculture by the ACI, which to this day has shown little concern for the hazards to man and the environment.

10.2 Concluding

In international agricultural research we have seen that developments are now aimed more at the situation of the farmers, because there is a growing recognition of their knowledge and experience. The international research institutes, however, do not concern themselves with putting new development within the reach of farmers, leaving this to a large part to national institutes or universities. They take on a supporting role as far as knowledge and expertise is concerned.

The FAO seeks a much clearer link with the situation of the farmers. Extension programmes are being set up which very clearly have a supportive and also a directing function. Through this, and also through the "Code of Conduct" established by FAO in the area of pesticides, FAO exerts influence at political level.

Through its collaboration with national governments the ACI gains at least as much influence on the policy concerning pesticides as does the FAO, although with a completely different aim: the increase in pesticide sales. Within the ACI there is still little sense of responsibility concerning man and environment.

11. Research and extension at national level

As was discussed in the previous chapter, international research organizations mainly limit themselves to fundamental and supportive research. The exact application and practical value of new technologies has to be developed in each country and adapted to each region. Local even means that for each farmer the situation which requires the use of a new technology may be different. It is especially in this testing of practical values that a close interrelationship between formal and informal research is essential. Universities and national research institutes are responsible for the testing of new technologies and for research into the necessary adaptations; farmers form the user group, which must supply information on the applicability of a new technology. Their informal experimental research forms the basis of this. Extension here fulfills the function of an intermediary, who must ensure that the information is exchanged mutually between the two parties.

In most Third World countries either the importance of crop protection is not acknowledged or there is lack of funds. This leads to a defective "crop protection infrastructure" such as plant quarantine facilities, pesticide control laboratories and research laboratories.

11.1 Research

11.1.1 National research institutes

National research institutes have, partly, arisen from the former colonial research stations, and besides new ones have been set up by development organizations. The diversity in the various institutes is high, as they have often been established with little coherence with those already in existence. In contrast to the international research institutes, little is known about them, because not much is published in internationally known and recognized channels.

National institutes form an important link for the international institutes as far as the application of results is concerned and they have, in principle, an important national task to fulfill. They have to work under such working conditions that their achievements sometimes are disappointing. On the whole, there are problems such as:

- A chronic shortage of provisions as far as finance, transport, research and library facilities are concerned. Agricultural research is often a neglected branch in agricultural ministries in the Third World. In the Third World more money is spent on extension than on research (in contrary to in the West). Priority is directly given to agricultural production and not so much to the support services such as research. Staff is also often badly paid through lack of finance, and can thus be poorly motivated. This can lead to the situation that they make use of every opportunity to get a better job either overseas or in the commercial sector (the so-called "brain drain"). Lack of money also limits the possibilities to issue publications and for main-

THIS NEW VARIETY IS RESISTANT TO STEM BORERS, BUT NOT TO SHEATH BLIGHT

Figure 11.1. At national research institutes ideas of international institutes are worked out for the national situation, keeping in mind farmer's practices.

Figure 11.2. National research institutes are dependent on the government, due to lack of funds staff is badly paid and there is lack of facilities. Due to these problems it is difficult to perform sound long-term reseaech.

taining contacts. This creates many bottlenecks in the attempts of such institutes to keep up to date with new agricultural developments.
• A crippling bureaucracy and a frequently very hierarchical power structure. Such a top-down approach, together with frequent transfer of staff, forms a great obstacle to good quality research within the institute and to maintain contacts with other services.
• The absence of clear objectives, which all to often results in research that is not very relevant, with out-of-date techniques and methods. The quality of the research often leaves much to be desired, especially as there is often no room for feedback from farmers during or after the research.

This view of the situation certainly does not apply to all, or to all parts of national research institutes. Moreover, the research capacity as well as the quality of research in the Third World is improving steadily. National institutes in Honduras, Guatemala and Costa Rica have, for example, produced varieties of pulses which have later been included in the tests by an international research institute (CIAT). It is clear that one can often obtain relevant and interesting information from the national institutes.

11.1.2 Universities
Many faculties of agriculture in Third World countries conduct research on crop protection. Sometimes this is fundamental research, but usually it is applied research, either with or without the collaboration of a national or international research institute. Sometimes, there are also studies on the transfer of technologies or research into extension methods. In most cases, however, there is hardly a link with the small-scale farmers. This is caused by the minimal collaboration between the national agricultural extension services and universities. It therefore often happens that new techniques or applications, which have been developed at the university do not find their way to the farmer.

11.1.3 Knowledge and experience of farmers
Knowledge of farmers is highly dependent on the region and personal development of the farmer. On one hand, farmers always have a valuable traditional knowledge on crop protection. On the other hand, knowledge necessary for applying new techniques often is incomplete, unless farmers are educated well.
For chemical control farmers have resort to only two sources of information: the extension services and the traders. Naturally, the objective of the pesticide traders is to sell as much of their products as possible and consequently rarely indicate negative effects or point to possible alternatives. The information available usually amounts to a description of the method of application; more fundamental information is hardly touched on. Moreover, labels are often misunderstood.
Agricultural extension officers, who should give unbiased information to the farmers, usually are governed by an absolute faith in the benefits and efficiency of pesticides and regard possible alternatives as second-class technology. This is the reason why farmers are not generally aware that pesticides can be a hazard to health and environment.

Farmers are likely to have only a limited awareness of the life-cycle of a particular pest and the build-up of a pest population, although this knowledge is essential for effective control. Lack

92

of this type of systematic background knowledge makes it difficult for farmers to recognise pests, and consequently errors are made, like:

- the symptoms are recognised but not related to the damaging agent;
- a pest is recognised but not regarded as sufficiently damaging to take measures;
- the wrong organism is blamed for the damage, farmers frequently fail to recognize 'useful' organisms which act as natural enemies, sometimes even blaming these for the damage caused by pests.

However, all this does not mean that farmers always fail in identifying pests. Cases are known in which farmers have correctly assumed that a certain pest was not or only slightly harmful, or in which they have correctly recognized natural enemies. One incident which illustrates this is that of the "shoot-fly". Although numerous publications report that this insect is a major pest in cassava, farmers in the Dominican Republic do not pay much attention to the shoot-fly and claim it to be quite harmless. Research proved that the farmers were in the right.

The result is a lack of knowledge and skill necessary for the deliberate decision to apply a pesticide. The user should be able to answer the following questions:

- Which organisms are causing the damage?
- In which stage of the life cycle can it best be eliminated?
- Which pesticide should be used and what are the associated dangers?
- When is it economically justified to spray; i.e.: what is the damage threshold?
- Which climatic conditions promote the effectiveness of the pesticide?

Proper application of pesticides is also fairly complicated: the pesticides must be properly distributed over the whole crop, on the right spots within the crop. Finally, pesticides can only be effectively and safely sprayed if equipment is available.

The poor application of pesticides cannot be entirely blamed on a lack of knowledge or technique. The concept of modern chemical control relates only to a very limited degree to the actual everyday world of the farmer in Third World. As a consequence, the requirements for the correct use of pesticides do not relate to the traditional knowledge of agriculture and crop protection.

On the other hand, farmers often experiment with new technologies, adapting these to suit their own particular circumstances. The realization that farmers do experiment with and adapt crop protection methods and that this is in fact highly desirable, clearly indicates that "package-deals" are not desirable at all. If pesticides are only available in pre-measured amounts (e.g. for a whole acre) in combination with artificial fertilizers and seed, the farmers are left with little room for their own experimentation.

11.2 Extension

For a long time, extension was the step which followed after research. This seems logical: research finds a solution to a certain agricultural problem, and must find a way to transfer this solution to farmers who then put the new technology into practice. The way to transfer knowledge were the extension services. This is schematically shown in the following diagram:

Figure 11.3. Extension officers translate the message of scientists to rules of thumb for the farmers, and should pick up comments of farmers for scientists.

RESEARCH ────────→ EXTENSION ────────→ UTILIZATION

Nowadays, it is recognized that this one-way flow of information, i.e. from research through extension to the farmers is insufficient. Extensive interaction between the different components is essential to obtain satisfactory results. In this, the extension services play an essential role: on one hand by taking information from one party to another (i.e. from research to farmers), on the other hand by effecting the necessary feed-back between the two parties. In this sense, extension is no longer the following step after research, but can take place simultaneously with, or even prior to it. The diagram becomes then as follows:

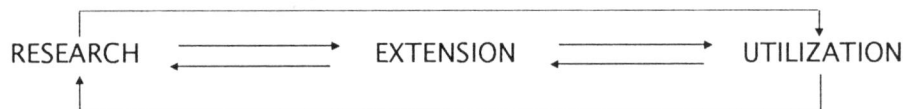

RESEARCH ⇄ EXTENSION ⇄ UTILIZATION

The arrows going left represent the process of "problem definition", the arrows to the right the process of "problem solution".

Research into crop protection can be at a fundamental level (studying the mechanisms responsible for resistance) or it can be highly applied (searching for methods to control the yellow rice borer in Thailand). Agricultural extension is mainly attached to applied research since its function is to transfer new technologies. The more research has an applied character, the more research and extension converge on each others' specific fields; sometimes to such an extent that they are virtually no longer distinguishable.

11.2.1 National agricultural extension services

The extension services generally is a service of the ministry of agriculture. Mostly, the structure is hierarchal with the village extension officer at the basis of the pyramid. The extension officers are rarely responsible solely for crop protection; in general they are responsible for extension on the whole growing process of several crops. The information which these officers take is often determined nationally, sometimes regionally. The current practice is to recommend the complete "package" (seed, fertilizers, pesticides) and credit facilities. However, it is often far from certain that the recommended inputs will be available at the time when they are required. Transport especially is often a major problem and certainly in vast agricultural areas like in Africa.

Extension officers promote the recommended package by means of demonstration fields and by visiting and advising individual farmers. In addition, they organize seminars, courses (usually on a district level) and sometimes they spread information through newspapers or radio. Very often a number of farmers are specially selected to act as "demonstration farmers", assuming that other farmers will automatically take over the new methods on their farms (see also chapter 18). The success of an extension service is usually measured in terms of production increase in a region; rarely thought is given as to who is actually producing more and how this is achieved.

In spite of the high investments in extension relative to research, there is still a serious shortage of funds. This means that personnel (in particular the village extension officers) are badly paid and do not have sufficient means to conduct their work satisfactorily. Moreover, it occurs frequently that bureaucratic and administrative work keeps extension officers from getting down to the actual extension task. Extension service are also sometimes called upon to assist with the provision of inputs, and they may also be involved in purchasing agricultural products where this is conducted through state cooperatives.

The greatest problem, however, is the lack of information relevant to farmers which makes it very difficult to take a realistic message (and to ensure feed-back).

11.3 Limitations of extension and research

Researchers tend to overestimate the capacity of the extension services as well as their access to sources of information. The application of IPM more than demonstrates this. Naturally, the components of the IPM-system dictate how much is required from the extension services. It is easier to inform farmers about the cultivation of resistant varieties or the application of other cultural practices than, for instance, about the application of damage thresholds and proper spraying techniques. The latter requires the ability to identify pest organisms and natural enemies, to assess damage levels, to know which is a correct pesticide to use and to be able to apply it in the proper way. Usually official recommendations are not at all comprehensive to farmers. They are formulated under the assumption that an extension service understands them and informs all farmers.

It is obvious that it is very difficult for extension services to ensure that the information does reach all farmers. Richer farmers usually have more contact with locally important people like extension officers, as well as with creditors and traders, and they have more possibilities to experiment new techniques.

For IPM also it is important that pesticides are available at the right time. The responsibility for controlling the quality of pesticides is a task of the national governments, leaving farmers at the mercy of a control susceptible to fraud.

Surveillance and early warning systems in crops sometimes also form a part of IPM. Research institutes often leave the responsibility for this to government services, although a lack of funds frequently means that these are incapable of fulfilling such a function, and the farmer is once again the dupe. To tackle this problem, agricultural research must not view the limited infrastructure as an obstacle but as a starting point for further research.

In the IPM programme in the Philippines, discussed earlier, so-called "pest-scouts" are trained to cope with this problem. A number of trainees in each village are taught how to detect pest insects. For example each week, they count the number of insects and damaged plants in a rice field. On the basis of these data and their knowledge of damage thresholds, they advise farmers on the best time for spraying. The money the farmers save in this way, can then (partly) be used to pay for these low-cost pest-scouts.

11.3.1 Criticisms

Conventional agricultural research is currently under criticism from several sides and new tendencies can be observed. It is interesting to notice that nowadays the research is the predominant target of criticism instead of the extension services. There is an ever growing

Figure 11.4. Local extension officers in general have a difficult task to spread information on the proper use of pesticides. It is difficult to get unbiased information and information on realistic alternatives, which can be accepted easily by the farmers.

IF YOU SEE THE YELLOW DWARF APHID YOU SPRAY APHOCIDINOL

IF YOU LOOK WELL YOU CAN SEE IT IS NOT THE YELLOW BUT THE ORANGE DWARF APHID AND BESIDES I CANNOT AFFORD APHOCIDINOL

Figure 11.5. Extension officers and farmers can learn a lot from each other.

conviction that much of the research conducted and many of the techniques introduced are of little relevance to small-scale farmers. It is more and more recognized that a number of biases of scientific research are partly responsible for this, and serious attempts are being made to take this into account. The most significant attempts will be discussed in the following.

11.3.2 Criteria for assessing the quality of the research

One of the obstacles for crop protection research relevant to small-scale farmers is the unsuitability of some of the elements of formal research. The orderly test fields at the institute, where all factors can be regulated at will and where reliable results are ensured, are not sufficient. They must be supplemented by on-farm experiments. Such experiments are much less controllable and have much more variable factors. Both forms of research are, however, necessary for a proper development of new methods.

Conducting on farms experiments is, however, not attractive from a scientific point of view and even less so because an important goal of research has always been the publishing of articles in scientific journals, this being a means to measure the quality of research. On site experiments are statistically very complicated and are far less conducive to the extraction of reliable scientific information.

Farming Systems Research (FSR) can be seen as a reaction on the mentioned criticisms and this approach is finding success in both international and national institutions. It focusses on the farm household and views the farmers' knowledge and experience as an important source of information. FSR is conducted by interdisciplinary teams to ensure that all aspects (technical and social) of a farming system can be studied. Much emphasis is placed on the practical objectives of this approach. A plan for FSR must also include steps for informing farmers about the conclusions. This is why the method is also called "farming systems research and extension". Crop protection can never be more than a small part of FSR and cannot be viewed as different from other bottlenecks, but since in most small-scale farming systems crop protection is an integral part of the whole system, this approach can be well linked up with the approach of farmers.

Farming systems research and crop protection.

The Asian cropping systems working group in cooperation with IRRI has developed a procedure for arriving at crop protection recommendations using farming systems research methods. The procedure consists of the following stages:
1. Understanding farmers' current insect control practices and resources available for insect control;
2. Determining yield losses for each crop growth stage;
3. Matching key pests to measured yields;
4. Selecting appropriate pest-control technology;
5. Testing the technology on farmers' fields in cropping systems managed under stable conditions over several years;
6. Evaluating the costs and returns of the technology.

Obviously before farmers can be informed, the possibility of adapting the outcomes to fit in with the whole system must be investigated which for the more complex farming systems is a complicated procedure.

Although farming systems research apparently obviates many of the shortcomings of conventional research, it also deserves its share of criticism. The practical application of FSR brings a number of weak points to light. For example, although in theory the participation of farmers is greatly stressed, in practice it is seldom carried out and methods are not worked out to bring this about. A second point is that FSR starts from the principle of multidisciplinary cooperation. In practice this is also problematic and in fact the various parties tend to work alongside each other rather than together. Thirdly, the amount of data which must be collected and analyzed before a clear picture of the total system is obtained is considerable. This is risky for both the clarity and the usefulness of the data and in fact, for the feasibility of the whole project. A fourth shortcoming is that the extension phase is not further developed, and usually entirely left to the extension services. Finally, in FSR too little account is taken of the fact that farming systems are continually changing and an analysis of the system remains valid for only a limited period of time.

In spite of this criticism, FSR is an important step in giving consideration to technologies which are relevant for small-scale farmers, and for the insight it gives into their situation.

12. Extension for small scale farmers

IPM can be regarded as an alternative to the conventional chemical crop protection and is available for both the traditional and the "modern" farmer. In general farmers are interested in stability in yield, lower costs and health risks.[3] An IPM programme must pursue and integrate these objectives so that it becomes a coherent and attractive programme for farmers. Of course, the prevention of pests exceeding the damage thresholds plays an important role in such a programme.

This chapter discusses extension methods for IPM, which take into account views and experiences of farmers. Using the FAO IPM-programme in Asia as an example the various stages of an extension strategy are elaborated.

12.1 Understanding the situation of farmers

Even if IPM proves to be feasible and seems lucrative on a research institute, this does not necessarily mean that it will work as effectively on a farm. On one hand this is due to the significant influence the more favourable environmental factors of the research institute (better soil conditions, irrigation facilities, availability of fertilizers, pesticides and expertise) have on the research results. On the other hand, however, it is important to realize that a farm is a system: all tasks performed by a farmer or family, including crop protection practices, fit into the system of that farm. Small-scale farmers in particular have few possibilities to control environmental factors. For this reason recommendations must be tuned to the specific circumstances of each farm. The reduction in yield of one crop caused by particular weeds for instance may be acceptable to a farmer if the weeds can serve as vegetables. Other, non-agricultural activities may also form a part: small-scale farmers frequently have off-farm activities competing with IPM (or agricultural activities in general) with regards to the available time, energy and capital. Minimalizing risks is an important motivation for small-scale farmers not gratuitously adopting new techniques: one failed harvest means the family will starve or repayments cannot be made. If a new technology brings an increase in risks and if it threatens to compete with other crops for labour or other inputs there is little chance of adoption.

However, not all farmers are alike, as was made clear by the effects of the green revolution. The larger farmers were able to profit more from the new technologies and through this they were able to oust the smaller farmers. Even now usually richer farmers see more chance of utilizing the various services and new technologies while these remain out of reach of the poorer ones. Small-scale farmers are unable to take the risks associated with a new variety, they do not have the money to buy pesticides, are unable to obtain credit and are not reached by the extension services.

An IPM programme must also take into account that in agriculture there is a traditional division of tasks between men and women. According to the separate situations and to the

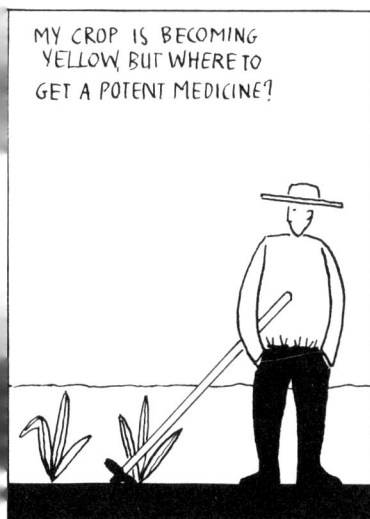

MY CROP IS BECOMING YELLOW, BUT WHERE TO GET A POTENT MEDICINE?

Figure 12.1. Farmers apply practical rules in order to get their crop grown well.

Example

In Nigeria an economical method was developed for protecting cowpea from bruchid damage after harvesting. The cowpeas had to be threshed and stored in cotton-lined polyethylene storage bags. This method, however, has never been applied: as soon as the cowpeas had been harvested, the other crops had to be harvested also and so there was no time for threshing.

socio-cultural environment, it can then be decided whether men and women should be approached together or separately. For example, how often did it not happen that a group of men was advised on how to prevent damage to stored products by insects by treating the harvest, while harvesting appeared to be a task for women? Often more intensive weeding is recommended ignoring the fact that this is a task of women and such recommendations will only mean an increase to their already heavy work load. Attempt should be made to estimate the effects of the introduction of IPM on the task division between men and women.

The heterogeneity of the agricultural population and the fact that the introduction of new technologies can increase inequality between different groups of farmers are now generally recognized. However, it remains very difficult to turn the pious wish to "develop technologies that benefit the rural poor" into concrete actions. Attempts are certainly made, such as packaging of inputs and seeds in quantities that are suitable for mini holders; focusing on small-scale farmer commodities and kitchen garden crops; not introducing technologies that diminish rural employment; paying attention to low external input technologies. IMP, for example, could be a technology which benefits small-scale farmers, because it costs less to apply because pesticides are partly replaced by labour like monitoring pests. However, this also does not necessarily hold true: is the information which the application of IPM requires available to small-scale farmers? What is workable for small-scale farmers when applying IPM? What is the cost of the extra labour that this technology demands?

12.2 How farmers perceive pests

The results of research conducted among some hundred farmers in the Philippines and Malaysia indicate that rice farmers perceive losses from pests as intolerable and inevitable unless regular preventive actions are taken. They estimate losses between 20% and 100%. They regard pesticides as their only way to ensure the yield, and are prevented from regular treatments only by the high costs of pesticides. They perceive pest damage as potentially within human control and pest management not as following rules of investment and economic return, but rather as an unavoidable part of modern rice production. The use of pesticides is "modern" and guarantees high yields and profits.

However, the notion held by the farmers that pesticides are essential to guarantee a reasonable yield does not hold up against various observations in the field. In half of the total number of fields observed it appeared that insecticides had no beneficial effect on the yield. No definite explanation can be given for this conflict between the belief of the farmers and the results of the research. Farmers can often misconstrue the cause of damage to their fields and then spray with an ineffective pesticide. Because farmers have no objective criteria to determine the effect of a treatment on the yield, they may conclude that the yield has been saved while in fact no positive effects have been obtained at all. Furthermore, farmers tend to spray preventive rather than curative. They are also apprehensive of applying simple damage thresholds, because under certain conditions, such as multiple infections or the threat of a virus, even a slight infection can cause considerable damage to a crop.

Of course, the attitude of farmers is highly dependent on the local situation. In remote areas in the Himalaya of Nepal for example, where little modern inputs are introduced yet, farmers see losses from pests as necessarily occurring. They know pests can be treated with "medicine", but nobody expects pesticides to be within reach for large scale application. Other constraining factors (like water supply) are much more in the picture.

12.3 Cooperation with small-scale farmers

A chapter dealing with extension and education should make clear what constitutes good extension and how this can be implemented in practice. We feel that a starting point for good extension is cooperation from (as much as possible) equal positions. An important question is: how can a field worker clarify something to the people he/she is working with? There is obviously not one general answer to this question. However, what should be evident is that the method for developing knowledge and informing farmers greatly influences what they are able to do with it.

Authoritarian extension methods, for example, give low returns (little knowledge is retained) and overlook the fact that farmers must be able to improve their crop protection techniques independent of "experts".

The objective of good extension must be to improve the farmers' grasp of their own situations. Improving crop protection must be through exchange of knowledge and experience between farmers and extension officers. Farmers should have a say in selecting and analyzing the problems and in deciding the problem approach.

12.3.1 Assessment of the level of cooperation

It often appears that farmers have no control on the process of cooperation for example due to institutional restrictions. For this reason we mention here questions by which the influence of farmers in a project can be determined.

Target group. Which groups of farmers actually profit by the cooperation? In regions with large social differences (Latin America and Asia) richer farmers may try to manipulate cooperation in their favour, while small-scale farmers are not likely to inform outsiders about this unless a situation of trust has been established. An extension officer must be aware of this possibility. In addition, one should pay attention to the accessibility of the information for the small-scale farmer.

Signalling problems. Who decides what problem is to be dealt with? Does the researcher mine the greatest bottle neck in a region by studying survey results, or do the farmers th selves determine what problem is effecting them most? And of course, again the questic "which group of farmers is affected by what problems" must be posed.

Organization of cooperation. How does the process of cooperation work? How much does it offer for contributions from the farmers? During the cooperation an extension can function as an expert who provides the solutions to problems the farmers encoun also possible that through discussions farmers are stimulated to find their own solutio example by conducting their own field tests.

How intensive should the cooperation be and who determines this? Is information al cient, or is a greater contribution from the extension officer required? Do the farmer organizing, or is it the field worker who finds this organization necessary?

Problem solutions. Who decides in which direction the solution should be sought? example, it could be decided to adhere to traditional crop protection techniques, to

more western approach, or to apply a combination of both. May a solution demand relatively much labour, time or money? In short, who decides the criteria for assessing the solutions?
Transfer of information. Who determines how the information is to be put across; the method of research; the structure for possible extra training?
Evaluation. Who decides on the most suitable form for evaluating the cooperation and its results for the farmers?
On the basis of these points a field worker or extension officer can make a structured assessment to what extent farmers actually are able to have their say in the cooperation.

Example

In West Java a number of years ago, there was the continual problem that fish bred in rice fields were killed by pesticides, which was recognized by the inhabitants. It was not clear whether they knew exactly how this worked, as Javanese people tend to be hesitant about telling about their experiences and speak in extremely veiled terms. Many people worked as agricultural labourers for large land owners. The land owners decided to apply pesticides in rice fields where labourers had bred fish for consumption.
Landowners may interpret criticism of the use of pesticides as being criticism to themselves. A field worker decided to use a Wayang Golek, a sort of puppet show, player to increase the people's knowledge about the effects of pesticides, at the same time protesting to the landowner at his decision to apply them. In the puppet show the gods declared their disapproval of the use of pesticides and apes were shown vomiting, which was immediately recognizable for the spectators from practical experience. Later it appeared that the authorities had also understood the message.
The man who gave the performance was by tradition allowed to broach sensitive subjects. Throughout the region in which he traveled he listened to people's problems and adapted his puppet show to present these. He himself thought the use of pesticides a problem important and fitting enough to be added to the several subjects he played about. Although the problem was new to him, this caused no difficulty, in fact it was part of his function to initiate discussions.

12.3.2 Forms of informing.

Language is, of course, of major importance for informing. We refer here not only to the language itself, but also to the symbolism implicit in its terminology. Moreover, the level of abstract thought must correspond to that of the target group. It is often necessary to demonstrate the topic in question as concrete as possible, i.e.: the preliminary measures which must be undertaken prior to spraying. If the subject does not lend itself to practical demonstrations then visual media can be very useful, such as posters, drawings, cartoons, slides and films.
These all make informative meetings more attractive. In Latin America, for example, the picture romance (a sort of comic with the drawings replaced by photographs) is popular. There they are likely to have more effect than a dry information booklet. Moreover, a large number of farmers are either illiterate or have difficulties with reading.
It is not advisable to show large blow-ups of pest organisms as these will not be recognizable to people who are not used to this.
Theatre and puppet shows seem to work well as a medium for putting across and spreading information. Even in regions where it has never been used, it can still integrate well with the oral and visual traditions. Radio and television are of course very suitable media for the spread of information on a large scale. Whatever means are used for informing, it is always important that this fits in with the local traditions.

're 12.2. Transmitting a message. Usually, there are local ways 'ormation exchange via which 'ation can be spread. Puppet story tellers, theatre, etc. useful to warn people 'ad consequences of pesticide use.

12.4 An example of an IPM-extension programme.

The FAO is running an IPM programme on rice in a number of Asiatic countries. The programme is implemented differently in various countries, basically following a scheme of five stages.

1. Raising awareness. The first stage is to stimulate a more critical attitude towards the chemical control of pests. Contact is sought with the village authorities and public meetings are organized to awaken interest in IPM. The meetings begin with a summary of the general problems experienced by farmers in the conventional use of pesticides.(Some of the frequently mentioned problems are: people being poisoned, the disappearance of fish from the rice fields, and the increasing costs of crop control.)

In order to gain an insight into the crop protection practices currently used by the village, a base-line study is conducted. The processed results are then relayed to the villagers and serve as the basis for further discussions. As a result of this they begin to regard pesticides as an investment with both good and bad consequences which can be weighed off against each other, instead of being a "must" for modern cultivation; a fundamental change in their attitude towards chemical crop protection.

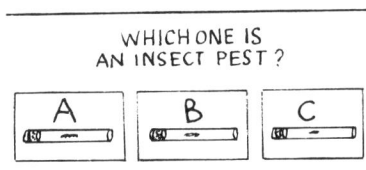

Figure 12.3ab. How to make farmers recognize pest organisms? One method is to offer them pest organisms in spirits and ask related questions; or make them collect "good bugs" and "bad bugs" separately.

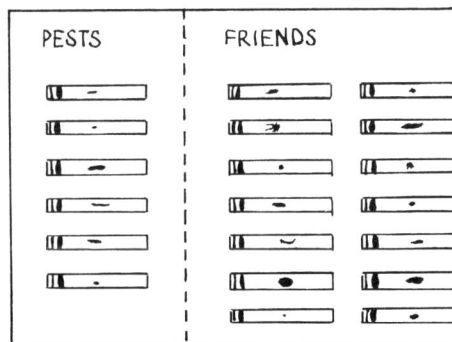

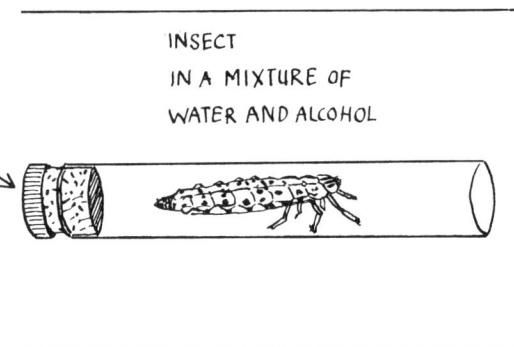

2. Explanations. After the introductory discussions, instructors organize a number of meetings with the farmers when crops are present in the fields. At these meetings farmers learn to observe pests and symptoms in actual field conditions. The explanations given during these meetings emphasize why IPM works better than the current chemical crop protection.) By discussing the ideas of farmers on the causes of a particular pest during the meetings, the instructors can adjust the extension message to local terms, measures and weights, practices and cultivation systems.
An important part of the training is the "scar-

ing session" which covers all negative consequences of pesticides. The participants become more aware of the character of these chemicals: pesticides are poisonous; they do not cure but are actually designed to kill. In many languages the same word is used for pesticides as for medicine, leading to misunderstandings with all the harmful results which this can bring. The risks to their own health makes as a rule the greatest impression on the farmers. In the Philippines the males in the groups usually suffered the greatest shock when they were informed of the vulnerability of the scrotum and of the risk of sterility.

Figure 12.4. IPM-training in the Philippines. Farmers are asked to mark three plants that should be sprayed, and three plants that should not be sprayed, under supervision of a trainer.

3. Field methods and demonstrations. Practical lessons are essential during the training. The ability to recognize pest organisms, symptoms and natural enemies is of great importance to IPM and therefore this aspect is given the greatest attention. It is quite a revelation for many farmers to realize that most of the insects they see are a help rather than a hindrance. Natural enemies are therefore referred to as "friends of the farmer".

Field training is also given to familiarize farmers with IPM methods, such as assessing damage thresholds. By imitating the instructors farmers learn skills in the fields. The instructors are at hand to point out any mistakes that are made. Once a farmer is able to identify the most general problems on his own, he/she can start comparing different pest levels and amount of damage resulting from the infestation. As far as possible assessment criteria are compiled from the situations observed in the fields. The more frequently one particular method for decision-making is repeated, the more the farmer will learn to trust the new techniques and this will only serve to increase acceptance of the method.

4. Problem sharing and discussion. Problem sharing between farmers, both in discussions and in field exercises, stimulates and establishes the IPM message by showing effectiveness of IPM in an existing number of local situations. Farmers are by tradition used to an intensive social interaction. As an organized group they are in a better position to point out to an instructor any part of new methods which is at odds with the farmers' existing practices. For this reason IPM can best be transferred within the framework of an existing farmers' organization, such as a village cooperative.

104

5. Learning and practicing including field tests. An essential part of the training concerns the evaluation method, which is aimed at improving the learned knowledge, to repeat and stimulate. Multiple choice field tests are used in the FAO programme to test identification of pests and natural enemies, determining possible damage caused by a particular pest, and choice of crop control method.
The tests are often carried out in the form of a game: a quiz in the class location or a ballot-box test in the field. For a ballot-box test approximately 30 boards are place in a rice field each containing a question. Three boxes are attached to each board on which is indicated three possible solutions. The participants walk past the boards and place a ballot (a piece of paper containing their name) into a box which according to them indicates the correct solution. Possible question are:

• Which insect is a pest organism or a "friend of the farmer"?

(in this instance there are three vials with insects attached to the board, and the participant must indicate the correct insect)

• Which leaf or which plant is affected by an insect?
• Is the damage in this field above, equal to, or below the damage threshold?

In the Philippines, such tests have proved to be very effective. However, it is possible that different results may be obtained in other cultures.

After such training farmers view the use of pesticides more as a financial investment than as a means to guarantee their harvest. They can also identify the most important pests and are able to decide whether spraying is necessary or not.

Farmers who have undergone training in IPM achieve higher yields and use less pesticides each season than untrained farmers. This evidence comes from the results of a survey conducted five years after the IPM training took place (see table 12.1).

Table 12.1. Rice production and use of insecticides by IPM-trained farmers and untrained farmers in the Philippines five or more years after training (averages of three crop seasons 1984-85).

	IPM-trained	Untrained
Yield (tons/hectare)	5.5	5.0
Insecticide applications per season	1.8	2.9
Costs of insect control (pesos/hectare)*	296	486

* $1 = 22 pesos (1985)

Altogether, it appears that with specific training it is possible to make IPM practicable for small-scale farmers. Group discussions, field demonstrations, videos and films are tried and tested methods to put farmers effectively in touch with their own situation and yet still allow them clearly to express their own points of view.

The success of an IPM programme depends on making the knowledge and experience of the farmers the central issue. It is also preferable that the programme is implemented in cooperation with an existing local organization capable of formulating problems and carrying out part of the research and extension activities. However, the presence of such an organization is more the exception than the rule and a whole process will have to be gone through to bring one into existence. Such a process forms part of the whole research and development programme. In this way the objective to "develop small-scale agriculture" changes to "teach small-scale farmers a process by which they can develop their own agriculture".

13. Conclusions

Apart from the international and national institutes mentioned in this part, there are, of course, many other organizations who concern themselves in one way or another with research into or extension on crop protection. It is impossible to review all these (see also part IV and appendix III). It is recommendable to make an inventory of what is done in the field of crop protection by the various organizations and projects occupied in a particular region of interest.

It is obvious that agriculture, including that in the Third World, has undergone stormy developments during the last three decades. In Asia especially, agricultural production has increased enormously. This development goes hand in hand with an enormous increase in poverty and problems of hunger. There is gradual realization that the challenge for agricultural research and extension is not only an increase in overall production levels.

The previous chapters described the attempts which have been and are still being made to make new techniques suitable for the small-scale farmer. One of the greatest problems associated with this is the complex system within which the agricultural activities of small-scale farmers take place and the huge diversity within these systems itself. Farming systems research is developed as a method for looking into these complex systems and from this knowledge for developing technologies which are acceptable to farmers within specific systems. Participation of farmers plays an important role in this but as it is still the researcher who determines the research subject and method, FSR does not yet yield much more knowledge than does conventional research.

Another matter which in the previous chapters was repeatedly pointed out as being a bottle neck, are problems of institutional nature: low prices for agricultural products, unreliable supply of inputs, inefficient extension services etc. In many cases the situation is such that no real improvements can be expected unless there is some change in this situation. Research cannot afford to ignore this problem, but unfortunately the claims of researchers that such problems are none of their business are frequently true, and were a researcher to take on these problems it would only increase the dependence on her/him.

Permanent changes to this situation can only be affected by the people themselves and researchers must be aware of this fact. These two points advocate that the essential parts of research, such as determining priorities, the diagnosis and the evaluation should be placed in the hands of the farmers themselves. Furthermore, the fine tuning of the developed technique should also be left to the farmers. This all looks well on paper but will not be quite so easy in practice. It requires that farmers are aware of their social position and that they have an adequate level of organization. In regions where this is lacking the priorities will always be to raise awareness and to form a farmers organization before a start is made with research or an IPM programme.

Part IV
Regulations and legislation regarding pesticides

Part IV looks at the economic considerations which play a role in pesticide production and distribution. It outlines existing international and national legislation. The expected effect from international regulations in the opposition to the misuse of pesticides is evaluated. The FAO 'International Code of Conduct on the Use and Distribution of Pesticides' is weighed in some depth as it contains much which could direct pesticide use and trade on to a proper path. Finally, the use of this information for the local development worker is discussed.

14. Pesticide trade and regulations

- INSUFFICIENT LEGISLATION
- LACK OF FUNDS, SO :
- LACK OF PERSONNEL,
- POOR LABORATORY EQUIPMENT,
- LITTLE KNOW-HOW

- POLITICAL BACKING, SO:
- WELL-WORKED OUT LEGISLATION
- ENOUGH MEANS
- WELL-TRAINED PERSONNEL
- WELL-EQUIPPED LABORATORY

Figure 14.1:. Who should control trade in pesticides?

14.1 Background

Up to now the correct use of pesticides has been central to this book. But it has already become clear behind the lines that many of the abusive practices related to pesticides make an early start at the production and marketing level. In spite of all the negative publicity (disasters, risks to health and environment), neither in the West nor in the Third World the use of pesticides is reduced.

The growth in the use of pesticides has given rise to a growing and blossoming trade. Policy and legislation with regard to pesticide production and trade has been a contentious subject since the beginning of the 1970s. In the West attempts to gain some control have had some success; especially for pesticides highly toxic or persistent, for example for the "drins" and DDT. The market of pesticides is quite impenetrable, for farmers and on the policy-making level, but why is this so? In part this can be attributed to the bad organization and inadequate legislation in a number of the Third World countries; and even in the West.

Multinationals have a particular role in this. The control of the pesticide market is namely in the hands of a few agro-chemical multinationals which have their home-bases in the West, the United States and Japan. These concerns can easily by-pass laws and legislation in one country by operating through their base in another. As this makes a considerable contribution to their profit margins, such practices are very common. National governments also often make concessions to stop activities being transferred to other countries.

This is one of the major reasons why in the last few years so much effort has been expended at international level to establish general guidelines for pesticide production and distribution. The frequent lack of legislation and control in the Third World has also been a factor, and, of course, it is now generally recognised that measures have to be taken in order to protect the environment and to achieve a sustainable form of agriculture. "The International Code Conduct on the Use and Distribution of Pesticides", an initiative of the Food and Agriculture Organization (FAO) an organ of the United Nations, is covered in this chapter in some depth as it contains a number of guidelines for the obligations of everyone dealing with pesticides; from the producers to the users. If this Code of Conduct were to become common practice, it would mean a great step forward to an improved, more open and responsible production, distribution and use of pesticides.

14.2 Regulation and legislation in the Third World

In many developing countries agriculture is seen as a "booster" to the economy. By intensifying agriculture, products become available for export and "hard" foreign currency is earned. At the same time it is hoped that there is also enough to cover the national food requirements.

To achieve this, Third World governments require access to the most modern technological products and production processes. This development has been fairly vigourous especially in Latin America and Asia, and one of the side effects has been an uncontrolled growth in the use of pesticides.

Insufficient regulation and control of pesticide production and use bears risks for the user, the environment and the consumer. This is the main reason for the need for good regulations.

A second important reason for establishing regulations is the ever tightening Western tolerance for pesticide residues in products imported from the Third World. Exporting countries must be ever more careful when applying pesticides at risk of losing their source of income. Morocco, for example, tightened its laws regarding pesticides after the EEC rejected various deliveries of vegetables because of high residue levels. Kenya, too, suffers from the same problem and is trying to introduce more effective legislation.

The third motive behind the call for improved control and regulations concerns the agro-technical problems rising when using pesticides (see Part II): some pests are becoming ever more resistent to pesticides, and "man-made pests" are being created.

Setting up an effective control system is difficult and very expensive; Third World countries often do not have the money, equipment, personnel or expertise to control the effectiveness of government regulations, quite apart from smuggling. Regulations therefore are not very well developed in most countries, and if they are, effective control is often lacking. In practice the pesticides market often is totally free, with all negative consequences.

For example, active ingredients and formulations which are not even registered in the West are being used in the Third World. Registration of a pesticide in the West demands a large number of tests. In the Third World many formulations are applied without having been extensively tested for characteristics such as persistency and toxicity levels, making it very difficult to keep an eye on the production site and methods and the export of many pesticides.

For effective control most Third World countries depend on the initiatives of the exporting countries and on international organs such as the United Nations organizations FAO and UNEP. For this reason, the Third World demands the imposition of export controls since the effective control of import is far more difficult to realise.

It is common for governments to act when serious economic losses have occurred. For example, the Indonesian government adopted IPM as a national policy after losing a total rice harvest to "hopperburn" caused by the brown plant hopper. Chemical control offered no solution to this pest as frequent spraying had already caused the disappearance of all natural enemies and allowed the brown plant hopper to multiply at will. The Indonesian government was confronted with enormous areas of infested rice and consequently with an enormous threat both to the national food supply and to export. Action was imperative. IPM appeared to

---

Situation regarding legislation in various continents

Here we describe the general situation regarding regulations on continents such as Africa, Asia and Latin America. These are intended to give a general impression of the situation in those areas of the world, but the situation may well show positive variations between individual countries.

112

be a good strategy for preventing hopperburn and it has since become part of the Indonesian agricultural policy.

14.3 Legislation regarding pesticides production and application in the West

At government level, in the West there is conflict between the short-term (economic) and the long-term (environmental) objectives.

The primary concern of the West is defending its own economic interests, and Western governments base their policies on pesticides on this premise. Accordingly, the "free-trade doctrine" which allows the consumer to freely choose from products on the market is very apt in that it shifts all responsibility on to the consumer of pesticides, i.e. the farmer.

However, considerations for the environment and health are giving rise in Western countries to the formulation of increasingly stringent legislation. Efforts are made to protect the domestic environment by restricting the use of certain pesticides, but at the same time the domestic economy and employment arouse an unwillingness to submit the chemical industry to anything more than the most minimal restraints. The remarkable result is that these restraints only apply to the **use** of the relevant pesticide in the country ordering the restraint and **not** to its **production** or its **export**. Shell Chemicals Netherlands, for example, was for many years the greatest exporter of Dieldrin although its use was actually banned in The Netherlands. This ambiguous attitude typifies most Western countries, and is referred to in this text as the "double standard". Out of all the pesticides exported in 1986, 30% were either banned in the country of production or its availability was subject to stringent regulations.

A note is appropriate here: a number of pesticides break down much faster in tropical conditions than in milder climates. In a few cases the reason for prohibiting the use of such a pesticide in the West (persistence in the environment and cumulating in food chains) is no longer valid in an importing Third World country.

Legislation in Africa

In Africa in particular, legislation regarding the use of pesticides is deficient. This has a number of causes. Before the 1980s, when international initiatives were established for regulating the distribution and production of pesticides, legislation in Africa was practically non-existent; the market in this area was totally free. The intensification of agriculture in Africa took place predominantly in large scale plantation agriculture and was usually governed by western industries. Thus, there was little incentive for African governments to introduce legislation through which in any case they would be unable to control practices on the plantations.

Joint interest associations, such as consumer groups and unions are also rare in Africa, especially in comparison to regions such as Latin America and South East Asia. In consequence, a vociferous cry for an improvement in conditions from the side of the user and consumer is absent.

Figure 14.2. The "double standard" of Western governments regarding pesticides production, trade and application.

14.3.1 New developments in Western policy

Today, a desire to safe-guard public health and the environment has convinced policy-makers in the West of the necessity to place restraints on the uncontrolled production and use of pesticides. The "double standard" is criticized more and more, and some countries are incorporating international guidelines in their national laws, like (mostly voluntary) variants of the "prior informed consent" principle. Within the present conservative/liberal governments and within industry there is a preference for agreements on a voluntary basis, so called "covenants", in order not to hinder free trade.

A number of countries in Western Europe have consequently increased the extent of their legislation.

Example

In The Netherlands in 1985 the Chemical Substances Act has passed. This has given the government considerable influence on the export of chemicals which could be hazardous to health or the environment. Before a product intended for export is loaded, the responsible government agency in the Netherlands and in the importing country must be informed, a type of "prior informed consent" procedure. If evidence warrants it, the Minister of the Environment may prohibit the export. The export control applies to 22 chemicals and participation is on a voluntary basis.

The Ministry of Agriculture released a long-term pest control note at the end of 1989. This plan sets out guidelines for structurally reducing agricultural dependence on pesticides within the near future. The number of pesticides approved are likely to be reduced by 50% (in the Netherlands a pesticide is forbidden unless it is approved). It is also probable that in the foreseeable future a levy will be placed on pesticides in order to finance research on alternatives, a practice that is already in force in Sweden.

Great Britain has a voluntary agreement between government and industry comparable to that in the Netherlands. The 1985 Food and Environment Act represents the legal basis for the regulation within Great Britain of pesticides and their export. The British government does not intend any direct involvement in export matters and its contribution is limited to one single notification to the importing country, and in this the voluntary cooperation of the exporters is anticipated.

Legislation in Latin America

Many Latin American countries do have legislation in the area of pesticides stemming from the forties when the level of development of most of these countries was comparable to that in Western Europe. One of the most important branches of the economy then was plantation agriculture controlled by large regional landowners and it was recognised that pesticides had undesirable side-effects. In general this gave rise to the establishment of laws to regulate the use and distribution of pesticides. Since the 1970s, due to the influx of pesticides intended to increase the yield particularly of cash crops and as a result of international discussions regarding pesticides, a number of countries have given their policies more incisiveness. Consumer organizations too contributed to this by making known the problems they had with pesticides and by expressing their disapproval. Often, however, there is a shortage of means needed to control the use and distribution of pesticides effectively.

14.3.2 EEC policy with regard to dangerous exports

The European Community (EEC) controls approximately two-thirds of the world pesticide market. Consequently it is of great importance that the battle to regulate the pesticide market has EEC backing. We shall therefore consider the discussions taking place within the Community, similar discussions are also taking place elsewhere.

Within the EEC the sale of a number of dangerous pesticides is either forbidden or is subject to restrictions. These measures, however, do not apply to export, and furthermore, the EEC via the European Development funds finances the export of banned or restricted pesticides. Originally the EEC did not wish to become involved in regulations of the export of pesticides: the import of chemicals was the responsibility of the importing country itself. Moreover, there was also concern that the competitive position of the European chemical industry against that of other pesticide exporters should not be endangered. However, political pressure from environmental groups and Third World countries has led to one sanction: "The Import and Export of specified dangerous pesticides act" which applies to 15 pesticides.

In contrast to other international regulations (from the FAO, etc.), EEC sanctions are legally binding for all the member states, which are internationally important steps forward in the area of pesticide legislation. Unfortunately, the EEC sanction is not very far-reaching and it is questionable whether it will be an effective instrument. The importance of the sanction lies essentially in its West European legal status and far less in the significance of its contents. In particular the double standard is retained.

After much internal contention, the EEC now recognises that the problems of chemicals must be tackled from within the framework of the general trade policies. Moreover, export should be based on the principle of "Prior Informed Consent" (see 14.6). However, the member states are less than enthusiastic about joining in such EEC sanctions. A number of them already have put some effort into incorporating the international guidelines of OESO, FAO or UNEP into their national policy and legislation, and still more legally binding legislation from the EEC would force the individual countries to adjust regulations once again. Apart from additional work, the EEC sanction could turn out to be less favourable to a country than its own legislation.

Legislation in South-East Asia

In countries such as Pakistan, India and Bangladesh, the earliest laws on pesticides also date from the 40s and were introduced after cases of poisoning and other undesirable side-effects appeared. In those days agriculture here was under influence by the West originating from the English colonial power.

Pesticides were initially introduced partly as crop protection and partly to eliminate vectors of human diseases such as the malaria mosquito and fleas. Recent developments have not led to these laws being given an edge and on the whole pesticide laws are only established or extended when there is an actual

outbreak of disease or when resistance develops. In South East Asia the people are relatively well organised, particularly among farmers such as the Farmer's Union in the Philippines, and in consumers organizations. These organizations wish better control of pesticides but they often lack any political power.

14.4 Pesticide production

There is a clearly discernable tendency to move the more dangerous production processes to countries with a less developed legislation, because environmental regulations in the West have become more stringent. Such moves are accompanied by the simultaneous transfer of the problems associated with pesticide production and distribution.

This transfer is a vivid illustration of a more general tendency for shifting "Western" problems on to those other areas, the export of chemical waste being another painful example. The dangers in the Third World are often much greater than in the West, because help at a sufficient level in emergency situations often is not available, disasters such as in Bhopal, India being the result.

14.5 Multinationals, the trade in pesticides and legislation

Not without reason, it is difficult to bring about an unambiguous and workable pesticide export policy. The role of the agro-chemical multinationals is decisive. Multinationals have establishments in numerous countries and this is the basis of their power. By observing in which countries certain products can be produced most cheaply, they are able to keep their prices far below that of their competitors bound to one location. The multinationals have their head-quarters as well as a few specialized production companies in the West from where they originate. Another characteristic of multinationals is that they aim to control the total production chain of a particular product. This gives them complete control over all price factors, and can increase profits to extraordinary high levels. Quite clearly, profit making is the driving force behind a multinational concern and it is with this motive in mind that they negotiate with governments on environment and health issues. A more responsible trade policy would be at the expense of the profit margins and for this reason the pesticide industry remains uninterested. In line with this, multinationals do not appear to cooperate in backing IPM by developing and producing more selective pesticides, because they are far less lucrative than broad spectrum pesticides.

As far as the Third World is concerned they point out the advantages of pesticides. An advertising campaign can easily communicate to a wide public that pesticides can have a rapid effect on pests. Multinationals have an enormous advertising budget and clearly their superior power in this area is great.

In economically weaker countries especially, multinationals have free-play. For example, if a country threatens to prohibit the production of certain pesticides they can be pressed upon with the counter-threat of cessation of local production, the consequences of which would be a transfer of production to countries where such prohibitions do not apply. In this way multinationals are able to exercise national and even international political influence. In the Third World they are important representatives of some Western countries of which they also make considerable contributions to the national income. All in all the multinationals largely determine which pesticides shall be available, when, where and at what price.

Worth a mention also, are the illegal pesticide manufacturers present in many Third World countries. Over this unregistered pesticide circuit, even multinationals have little control. Yet, multinationals have always opposed legal sanctions or legislation with regard to the production and sale of pesticides, because such legislation could undermine or hinder the control

Hans Traxler/ Welt der Arbeit

Do you have problems with the growth of your wheat? Simply spray 5ℓ of SUPEROL!

Unfortunately it will bend now, No problem, spray 10ℓ of STRONGOL

Now your wheat is growing upright. So are the weeds. Spray 50ℓ of SAMEX.

Figure 14.3. Your chemical consultant. (Pages 116 & 117.)

they have established over a production chain. They prefer to advocate voluntary agreements (covenants) with national governments. However, they make some effort in the area of extension on the safe use of pesticides (GIFAP pamphlets) and they cooperate in the transfer of technology to the Third World by providing Western manpower, offering training, etc.

Example

The following notice was published in "Onze Wereld" of May 1989. Whether the insecticide concerned is indeed such a "wonder-remedy" has not been investigated; it serves only to illustrate the influence of multinationals on national governments.

"There has been a recent scientific break-through in India: the researcher M.N. Sukhatme has succeeded in producing a vegetable-based, non-toxic insecticide known as Indiara. The Indian committee for the registration of pesticides refuses to include the new pesticide in its records, however, probably due to pressure from multinationals.

"Indiara is a so-called "contact-insecticide" containing, among others, garlic, ginger and mustard. Fifty per cent of all cockroaches, beetles, flies and mosquitoes coming into contact with the pesticide die within the hour. The remaining fifty per cent within a few hours. Indiara can be used for pest control and as a preservative for seeds and foodstuffs.

"It is many times less toxic than DDT and scientific experiments have revealed no negative effects on health after oral intake. The production process of Indiara is very clean and waste can be used for fertilization. Some months after being granted a provisional licence, Sukhatme received full approval for production from the Ministry of Agriculture. Subsequently, the pesticide was successfully applied on various fronts in the Vidharbha and Madya Pradesh regions. That is until the Central Insecticide Committee and the Committee for Registration informed Mr. Sukhatme in writing that his licence was being revoked. Sukhatme puts this down to pressure from multinationals; India spends 25 million dollars each year on the import of pesticides."

That's what the pests love! So what! Spray 100 l. of MILBEX.

In the extremely unlikely case your body should unfavorably react to the consumption of this wheat

don't worry: three daily injections of SANOSTRONG will do.

In Europe manufacturers of pesticides have laid down their reasoning in their own code and in fact reject all responsibility for their activities in the Third World. The central theme is that responsibility is a part of effective operational control; in other words that the host country should be responsible for the production of pesticides. Consequently, demands on the local governments become heavy. Based on the argument that safe technology will be introduced, the actual intention is to open up the local economy of the country concerned by admitting Western technology and manpower, a process which is stimulated by IMF.

In brief: agro-chemical multinationals are indeed willing to talk about reducing the misuse of pesticides both in the West and in the Third World. However, they are not prepared to limit the export of pesticides to the Third World or their production. For the time being they say that establishing voluntary based guidelines for unambiguous advertising and labelling will have sufficient effect.

14.6 International regulations

14.6.1 Inducement of international regulations

In 1977 at a conference of the environmental organization of the United Nations, UNEP, the Kenyan delegation for the first time within an international organization drew attention to the

117

dumping of pesticides in the Third World. Already at that time it was decided that the export of dangerous pesticides was not permissible without the prior consent of the authorities in the importing country. The principle was established that the governments of exporting countries should bear some responsibility concerning the export of dangerous pesticides, and here the provision of information and the awaiting approval were fundamental. This principle later came to be known as **Prior Informed Consent (PIC)**. Environmental organizations such as PAN have continually appealed for the introduction of PIC in international guidelines to avoid Third World countries becoming the victims of a lack of adequate national legislation.

Awaiting approval by the importing country proved to be a controversial political point. The exporting countries quickly took the lead in the discussion by offering their own alternative: importing countries should be given prior warning about restricted pesticides, but sanctions to prohibit export should be excluded. This alternative also only required a single communication per chemical and not one for each individual transaction. This agreement between exporters has set the tone for other debates at international conferences.

As a result of the unwillingness of multinationals and through pressure from the Third World and environmental organizations four leading international organizations (OESO, UNEP, FAO and EEC) established guidelines for the production and distribution of pesticides.

Voluntary notification of the pesticide to be exported (risks, toxicity etc.) is fundamental to all the guidelines but they differ from each other with regard to the recommended frequency and depth.

It should be noted that any guideline of this type applies only to a limited number of pesticides: UNEP has developed a list of 45 pesticides, the EEC sanction restricts itself to 15 pesticides, those that are forbidden in the EEC country itself. A list of the "Dirty Dozen" naming the twelve most toxic pesticides has also been compiled by environmental organizations.

Legislation from the OESO is the most decisive for what is undertaken on a global scale. This is largely due to the nature of OESO, which is ultimately a form of cooperation based purely on economics. For example, OESO considers a single notification stating the toxicity of a pesticide and whether or not it is prohibited to be sufficient for an importing country.

The **EEC-regulation** amounts to no more than an export regulation. An importing country is given one notification of a pesticide to be imported. If the importer then responds positively or not at all within a 60-90 day period, the export transaction takes place.

UNEP and FAO by contrast are on the point of including the principle of Prior Informed Consent as the effective legislation in their guidelines. Then, one single notification will no longer be adequate; the appropriate information must be supplied at each transaction or with each product. In terms of content, the UNEP and FAO guidelines are the most far-reaching. However, they are also the least radical in terms of effectiveness and impact as their application is voluntary. In the following paragraph, we look in more depth at the FAO Code of Conduct as it contains good starting points for a more responsible pesticide policy.

14.6.2 The FAO Code of Conduct

The FAO has developed the international Code of Conduct on the Distribution and Use of Pesticides with the aim of tackling the problems arising from inadequate pesticide legislation in the Third World. Non-governmental organizations (NGOs) and PAN (Pesticide Action Network) WHO, UNIDO, UNEP and UNESCO collaborated on the concept for the Code of Conduct. The pesticide industry was also represented in the shape of their interest organization

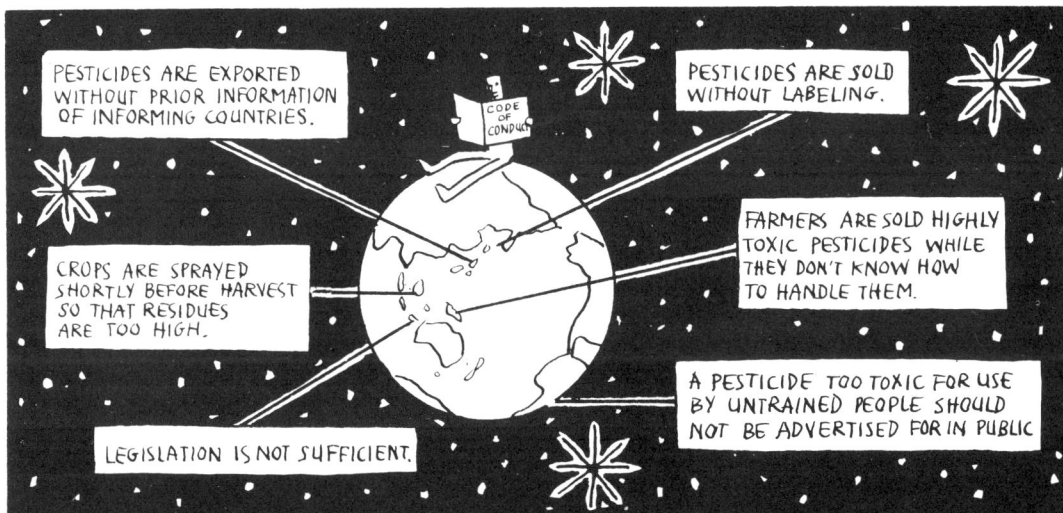

GIFAP. The Code of Conduct was accepted by a general vote at the end of 1985. Since then no effort has been spared in gaining wide recognition for the Code of Conduct. However, Third World countries especially, expressed doubt concerning the principle of Prior Consent, which was not explicitly included in it. Recent conferences indicate that this must take place by the end of 1989.

To give an impression of the contents of the Code of Conduct, the following is a brief summary of such of its articles which could be relevant to those involved in agriculture in a practical or advisory capacity. Articles concerning the responsibilities of manufacturers and governments are not included.

Article 1

sets out the objectives of the Code of Conduct which are described as being:
- defining the responsibilities and setting out the voluntary rules of conduct for all those involved in the production, distribution, transport and use of pesticides;
- underlining the necessity for collaboration between exporting and importing countries to ensure the safe and efficient use of pesticides;
- recommending responsible and universally applied trading practices, assisting countries which lack specific legislation, stimulating the effective use of pesticides with the aim of improving agricultural production while respecting the health of human-, animal- and plant-life;
- the Code of Conduct requires to be applied within the framework of national legislation as a basis by which governments, manufacturers, distributors and others involved can assess the general acceptability of their activities and those of others.

119

Article 5

concentrates on reducing health risks:
- establishing a pesticide registration and control system;
- the possible removal from the market of dangerous pesticides (usually determined from the WHO guidelines regarding dangerous pesticides, see also appendix I);
- providing advice on the treatment of poisoning;
- the separate storage of pesticides and food stuffs in shops where both are on sale;
- making available less poisonous pesticides;
- providing sound extension for the public.

Article 7

covers the availability and use of pesticides:
- it emphasises the importance of matching extension to the knowledge level of the user;
- pesticides should be classified according to degree of toxicity and very toxic pesticides should be prohibited unless sufficient guarantee of their safe use can be given.

Article 8

provides guidelines for the distribution and trade in pesticides:
- before being put in to use, pesticides should be tested in conditions comparable to those for which they are intended;
- quality requirements for exported pesticides should be the same as those applying in the country of export;
- pesticides should only be sold through bona fide channels and sale in non-approved packing must be prohibited.

Article 10

concerns the requirements for labelling, packing (both being the duty of the manufacturer), storage and treatment of waste. It is recommended here that the international guidelines developed and published by the FAO should be followed. These guidelines concern:
- efficacy data for the registration of pesticides;
- registration and control of pesticides;
- environmental criteria for the registration of pesticides;
- packing and storage of pesticides;
- good labelling practice for pesticides;
- disposal of waste pesticides and pesticide containers;
- crop residue data for pesticides.

Article 11

gives recommendations concerning the marketing of pesticides:
- misleading statements concerning safety during use and the effects of the pesticide may not be made;
- illustrations used in the information material must not depict any misapplication i.e. people spraying without sufficient protective clothing;
- pesticides which may only be applied by specially trained personnel may not be advertised for the general public.

Support for pesticide legislation in the Third World, and in Africa in particular, comes largely from the side of the FAO. In various regions the FAO Code of Conduct is taken as a basis for national legislation. Regions such as Central America, South East Asia and the Pacific have already held conferences on this matter. The earliest subsequent negotiations on the programme of the FAO are to take place with fourteen West African States taking part.
At present time the NGOs have the important task of evaluating the application of the Code of Conduct. A number of its passages more than once specifically address the general public; emphasising that everyone is affected by the pesticide problem and that we must all bear a share of the responsibility. Until now it is the only document which covers every aspect of the distribution and use of pesticides in great detail.

15. The Code of Conduct in practice

This chapter looks at the discrepancy between the FAO Code of Conduct and actual practice; in other words it considers what must still improve. An evaluation project was set up by the Environmental Liaison Centre (ELC, see appendix III) as a progress evaluation after the introduction of the Code of Conduct. Two years after the Code was accepted, ELC observed that in the thirteen countries evaluated the general practice with regard to the distribution and use of pesticides has hardly changed.

The overall attitude of industry as could be predicted is that the Code of Conduct works admirably and that it is observed. This is also the attitude of the pesticide producer organization GIFAP and a number of manufacturers' societies in Germany and Switzerland. Their members are supposed to take make an effort to follow the guidelines of the Code of Conduct. However, the evidence gathered by ELC from around the world suggests a different story: the efforts made by industry are far from adequate.

15.1 Labelling

To what extent does industry endeavour to ensure that their products are correctly labelled? The FAO guidelines give clear information about good labelling practices: labels should be clear, give adequate warnings, provide information about toxicity, state the WHO classification, be in the local language or carry pictures etc. In many countries the most simple and easily applied guidelines are ignored.

Example

In Ecuador an examination of 46 labels revealed that the majority did not comply with the guidelines of the Code of Conduct. Three-quarters of the labels had no serial number or production date. Almost half had a risk classification other than that of the WHO, more than a third lacked information about protective clothing and almost a quarter made no mention of safe storage or instructions for disposal. Ten labels failed to state the final day of use of the product before the harvest. Moreover, the condition of some containers was so bad that they leaked.

Frequently, pesticides are re-packed in smaller quantities to make them easier to handle. In such cases, good labelling, or indeed any labelling, was usually omitted. To avoid this type of abuse pesticides intended for retail should be packed directly in small packages.

Figure 15.1. A pesticide label, containing all information as required by the code of conduct. The label should be in the local language.

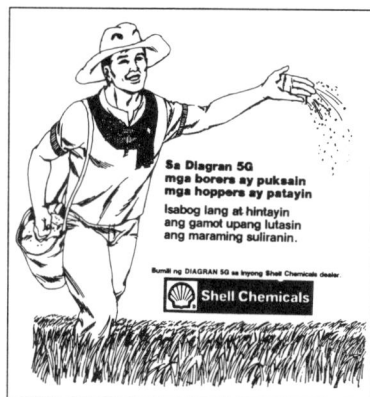

Figure 15.2. The code of conduct forbids pictures in advertisements where people apply the pesticide in an incorrect way. This man applies a pesticide without sufficient protection. Active ingredient of diagran is diazinon, an organophosphorous compound of which 20 grams is lethal for a normal human being.

15.2 Publicity

The guidelines of the FAO Code of Conduct concerning the publicity and the advertising of pesticides are as clear as those for labelling: information must be technically correct, it must contain no claims of safety, misleading promises or guarantees of higher yields and there must be adequate warnings about the use and toxicity of the pesticide, etc.

According to the ELC, producers all over the world are using advertisements and publicity material which totally ignore these guidelines. Very little of the publicity material provides the information needed for safe use. Advertisements generally fail to name the active ingredient, the composition-data nor the standard warnings. Rarely there is a note that the user should study the labelling carefully, and yet many companies claim to bring the very safest of products on to the market.

Example

In Brazil a newspaper for farmers described Decis (from the firm Roussel Uclaf, active ingredient: delta-methrin, a synthetic pyrethroide, usually chemicals which are not very toxic) as being "the safest insecticide in the world". Similar advertisements in subsequent numbers claimed that the use of Decis was "a safe decision". In Egypt and other countries in the Middle-East the journal "Middle East Agri-business" promised that with Decis "absolute safety is ensured not only for the consumer, but also for the manufacturer who is able to offer a product acceptable for world-wide export". In Papua New Guinea, Decis is described in a folder as being "extremely safe for man" while another Roussel Uclaf folder gives the impression that K-Othrine – deltamethrin under a different trade-name, has "an extremely high safety level". An Indonesian folder described Decis as "safe for man ... it leaves little residue". Deltamethrin in unfavourable conditions shows a LD50 for rats of 135 mg/kg, which means that 10 grams may be fatal to humans; besides it is toxic to water organisms.

Figure 15.3. Food stuff and pesticides should not be stored separately, and the latter should be sold in well-labelled packing. Moreover, when pesticides are sold protective clothing should be available.

Apart from exaggerated claims about the safety of a product, many companies make all kinds of promises in their advertisements: better crops, higher yields, more profit, and so on. For example an Union Carbide flier in Indonesia reveals that Temik (aldicarb) "increases the production of potatoes by 30%", a Union Carbide folder in Ecuador makes a similar claim about Temik.

Companies will then introduce still less valid advertising or publicity campaigns, for example in Thailand Shell issued coupons to people buying pesticides. The coupon is made up of two parts: the first offers discounts on future purchases; the second is a lottery ticket for a prize from a total prize range worth 1 million Bath (approx. US$ 75,000) and including gold, televisions etc.

Most of the publicity material collected during the ELC evaluation project failed to comply with at least one of the regulations laid down in the FAO Code of Conduct. Among the abundance of advertisements and promotion pamphlets collected in Asia, Latin America and the Pacific it is practically impossible to find one which satisfactorily complies with the FAO regulations.

15.3 Retail and use

The FAO Code of Conduct has appealed to industry "to be actively involved in keeping their product in sight until it reaches the ultimate user and to keep track of the main forms of its use. Any problems pointed out should be used as the basis for changes in labelling, instructions for use, packaging, formulation or availability".

In point of fact, no company can be held legally responsible for the trading practices of the middlemen, or for the practices the user fails to attend to. However, companies do have a moral obligation to tackle any dangerous and evil aspects of their products. According to the Code of Conduct, if the safe use of a product appears unattainable, the manufacturer should issue a sales-stop or withdraw products from the market.

In many countries sales-stops of some products should be issued since long times. There is virtually no evidence of a re-evaluation of pesticides other than for financial or commercial reasons and pesticides are only taken off the market when their commercial value ceases. Companies are reluctant to replace existing products by less toxic formulations, especially if they are less commercially attractive.

Example

In Egypt, in spite of the centralized stocking system, pesticides are freely available on the open market. As in many countries, there is virtually no control of the intermediate trade. The middlemen usually deal in pesticides, not in protective clothing, which is also not recommended to farmers who, in any case, can not afford to buy it. The middleman himself and fellow farmers remain their chief source of information.

15.4 Closing comments

The present state of affairs is certainly not a satisfactory one. The FAO Code of Conduct in spite of being a considerable step forward in international pesticide legislation must clearly be supported for more effects.

In the countries evaluated the laws controlling the safe use of pesticides are either antiquated or inadequate. Where there is any structure for market control a lack of means usually means that it is insufficiently supported.

The ELC report details the practical problems arousing when Third World countries carry out their responsibilities in the area of pesticide legislation and use. Third World countries need more support to overcome the gap between the code and practice, and technical and financial backing is needed to give support to national legislation in the developing countries. Regulations and laws are useless if they cannot be controlled. From the fate of the FAO Code of Conduct it can be seen that effective control has not been adequate. Self-control on the side of industry does not work, and NGOs only have limited facilities for carrying out fundamental controls and evaluations.

A worthwhile recommendation is therefore that evaluation guidelines be strengthened and that they contain specific responsibilities, for instance for the FAO who could act as counsellors in the event of infringement of the Code of Conduct. The FAO could also be given a mandate for installing an independent panel who would study data and make recommendations for any action to be taken. The support of legislation in the Third World is one such action. Furthermore, the FAO, international organizations and bilateral and other financing organizations, must be encouraged to contribute to a more systematic control, including making appropriate provisions available to NGOs.

16. Possible action

This book cannot provide any clear-cut receipts to cut the use of pesticides drastically. Awareness of the problems concerning chemical control in the Third World has only recently started to grow, and concrete solutions at farming level are thin on the ground. This chapter discusses possible actions for groups and individuals working in the area of development in the Third World and in the West.

The publication of Rachel Carsons' "Silent Spring" (1962) which deals with the effects of pesticides on people and the environment can be regarded as the starting point for increased public interest in the West in these problems. Since then, agricultural and environmental scientists, environment and Third World groups have been concerned with the effect on the environment and the export of dangerous pesticides to the Third World.

In a variety of circles the criticism of these developments has led to a deeper analysis of the virtually non-sustainable character of modern agriculture. Organic farming movements are gaining more supporters, and many groups are trying to move industries and governments towards environmentally sound techniques and better legislation, inter alia with regard to the pesticide problems.

Figure 16.1. Care should be taken not to contaminate (drinking-) water with pesticides.

Governments in most Third World countries are subjected to a complexity of development aid. As a consequence, there is state interference in many sectors of society but not of a structural kind; as it is generally coupled to assistance on project basis. When undertaking action in the field of pesticides, in many countries the government cannot be counted on. On the other hand more and more groups of people get conscious about effects of modern agriculture on public health and environment. A vivid example of this is the '88-'89 actions of the Indians for preservation of their living environment in the Amazon.

16.1 Environmental groups

NGOs like consumer organizations or environmental pressure groups can stimulate governments in undertaking action, for example by making accept the FAO code of conduct as a base for national legislation, or by making the public aware of the hazards of pesticides. There are a lot of environmental pressure organizations (like the World Wildlife Fund, Greenpeace, Friends of the Earth and Health Action International) which usually are not concerned with pesticides in particular. Other organizations are more active in the field of pesticide use and its alternatives, some of which are described below. Their adresses can be found in Appendix III.

16.1.1 Pesticides Action Network (PAN).

PAN was established in 1982 and is a network of organizations active in pesticides problems. More than 350 organizations in 42 countries are members, they exchange information and commonly undertake actions. One of these actions is the "dirty dozen campaign", which propagates withdrawal of some twelve most hazardous pesticides from markets, while better alternatives are suggested.

PAN participants have agreed to concentrate their efforts on achieving the following aims:
- the expansion of traditional, biological and integrated pest management and putting an end to the overuse and misuse of pesticides;
- the imposition of export and import controls on hazardous chemicals, in particular pesticides;
- immediate notification by governments of a ban or restriction on a pesticide;
- public release of information by governments on the export and import of pesticides, including the names of companies involved and the associated known/potential hazards;
- the withdrawal of funding by international development agencies of projects involving the use of pesticides which cannot safely used under local conditions;
- the reversal of the "green revolution" practice of developing seeds which need large doses of pesticides and fertilizers;
- putting an end to the vicious circle whereby pesticides used in the Third World end up as residues in food eaten all over the world.

One strong point of PAN is that it shows up with real facts and realistic examples concerning pesticides' use and misuse in the Third World, contributing considerably to the international discussions. PAN has an observer status at FAO debates and pleas for incorporating the Prior Informed Consent principle into the code of conduct.

16.1.2 Environment Liaison Centre (ELC).

ELC undertakes international efforts for establishing trade regulations, in particular the code of conduct. ELC is based in Nairobi, Kenya, being a regional coordination point of PAN. It is an organization of environmental and development groups. ELC has more than 250 members in 60 (mostly developing) countries all over the world. The main goal of ELC is to support NGOs in their efforts to stimulate implementation of sustainable agricultural methods. At this moment evaluation of the code of conduct is a main activity.

16.1.3 OXFAM.

OXFAM is a British organization supporting development in the Third World. It requests governments to establish effective control measures for pesticides, and stimulates application of IPM in Third World countries. OXFAM calls upon all parties involved to search for better control methods, from individual to mondial level. It requests EC to establish a good information procedure about import and export of pesticides. The manufacturers are requested to obey all kinds of codes of conduct, being voluntary or not. OXFAM has published a lot of brochures and books on development problems, starvation, medicines, including on pesticides.

16.1.4 International Coalition for Development Action (ICDA).

ICDA is concerned with (among others) pesticides problems and problems in the seed breeding sector. From this starting point ICDA found out there was a need for facilities for conserving and maintaining land races (gene banks). It pleas for new structures to bring about this, for example an international treaty. Agricultural research should be ameliorated and genetic manipulation as well as monopolizing genes should be controlled better.

ICDAs philosophy is that in principle Third World countries should be able to determine developments in their own country themselves. Direct contacts between Western and Third World groups and individuals are necessary for supporting one another. Groups in the Third World deliver information on problems and solutions, Western groups can use this information to clarify and make known what happens in other parts of the world.

16.2 Possible action by development workers.

What can a development worker do about responsible use of pesticides? As said, there is no clear-cut answer. We only can suggest some possible ways of action.

16.2.1 Providing information.

A development worker can be an intermediate between international or environmental organizations and projects in the Third World. Getting in touch with information about IPM in particular crops or about pesticides or extension material often is not very simple. Sometimes the national ministry of agriculture has some information available. Information of FAO is very valuable, providing own publications as well as information from WHO, IRRI, etc. This material often is available from the local office of FAO, or from book stores, ministries or embassies.

On the other hand there is the need for information in the West, for example about policy regarding pesticides in rural development projects. This information may be a base for action in Western countries.

16.2.2 Field contacts.

A very important source of information are direct contacts in the field. Via counterparts or

Figure 16.2. Development workers usually have more access to literature than local people. They also can be an intermediate between international environmental organizations and local initiatives.

informed persons Western field workers get an entrance to governmental structures. Abuses in the trade and use of pesticides can be signalized and passed on to authorities. The FAO code of conduct can be a good reference point in this respect.

Most Western field workers only stay a few years in a project and therefore it is of great importance the local staff supports the need for conscious use of pesticides. Only good collaboration of the field worker and local staff allows both parties to learn from and understand each other. Every situation apart determines the proper methods to achieve this.

A very promising method is searching contact with local cadre open to improvements and new unconventional views. Good contacts between this kind of people and field workers ensure ideas for alternatives to be worked out well, and they may support each other. Western development workers got a different influence in such combinations, via their different background and easier access to information. Without doubt, this kind of contacts can be very useful.

17. Conclusions

Due to the general threat to the environment and to public health, increased activity to regulate the use and distribution of pesticides can be noted. Consciousness of the risks associated with pesticides appears to be growing, in the area of policy-making as well as at the application level.

The international organizations are working hard to improve the existing import- and export regulations about pesticides as well as environmental aspects and the position of the Third World countries with regard to the production and trade in pesticides. Recent decisions of the UNEP and the FAO have been epoch-making in the introduction of Prior Informed Consent: the importing countries will have a much greater say in deciding what they want to import. Following up these regulations, however, is still hampered by many hindrances. On paper, everything appears to be better arranged and under greater control, but that is only a first step. Are similar steps at international level adequate to call a halt to the increasing use and misuse of pesticides? Regulations embodied in law, after all, still only exist within strict limits. Under the influence of industry, the endeavour has been to obey the regulations on a voluntary basis; there seems to have been an attempt at all costs to avoid the provision of sanctions in cases of misuse in pesticide production and trade.

The first studies to evaluate how the FAO Code of Conduct is followed in the Third World do not provide a rosy picture. Publicity about pesticides is mostly misleading or incomplete. The pesticides in use often belong to the most dangerous groups (WHO class 1A or 1B; see appendix I) and they are often transported, sold and used under unsafe conditions. Most governments of Third World countries lack the ability to control the large-scale and very profitable trade in such dangerous products or to provide the necessary information and training programs in order to stimulate safe usage.

Defective control and a lack of means are not the only factors that put a brake on the introduction of responsible pest control. Both in the Third World and the West, political and economic relations are decisive, and here many contradictory interests can be found: environment, profit, export, public health, employment etc. Therefore it is not surprising that there can be no real talk of sound regulation of the production, distribution and use of pesticides.

Under pressure of public opinion, Western governments are presently making their national environmental laws more effective. Therefore, industry is removing its production bases to the Third World where environmental rules are less stringent, and in this way the problem is simply moved on.

Last but not least, better co-ordination of the activities of the various international organizations would surely contribute to their effectiveness. Apart from the FAO, there are a number of international and UN organizations that concern themselves with pesticides in their own way and in their own area of activity. This ultimately leads to a confused mass of dissimilar authorities and thus an ineffective approach to the problem: just a single general authority is essential to make inroads on global problems such as that of pesticides. Nothing of the sort can be expected from the UN in the not-too-distant future, but there is more to hope from the EEC. Decisions by the Council of Ministers and of the Commission can immediately be included in the legislation of the member states, the only disadvantage being that by such legislation, measures can only be taken in Europe. However, this can act as an example to other countries.

At the international level, organizations such as PAN have achieved observer status at discussions, for example at the FAO Code of Conduct, and they have the ability to exert some pressure on the decisions which are taken there, although the effect of this is often limited.

Joining an organization or becoming an individual member of one of the groups which are members of PAN indicates support for their activities and indirectly also has an influence on international points of view regarding the trade, production and distribution of pesticides. The same can be said on a national level: to give national consumer and environmental organizations more influence, as many people as possible should join them.

Locally, the individual field-worker or extension officer can try to watch out for irresponsible use and distribution in pesticides. In this case also, membership of an environmental organization or of PAN can bring about a better exchange of information: to keep the field-worker informed about the dangers of certain pesticides and possible alternatives to them, and in other directions to pass information from the field to these organizations.

Despite all actions undertaken at various fronts the situation regarding pesticides use is not very rosy. Changes in positive direction and striving for more endurable agriculture just slowly get under way. This has political, economical and social causes, which are very difficult to change. There is a long way to go, in which a stimulating function of environmental movements is indispensible, both at national and international level.

Part V
Case studies

The last part of the book consists of two case studies, in which especially theories from parts II and III are elucidated and integrated. The first case elaborates on IPM in rice in Sri Lanka as part of the FAO IPM-programme in Asia. A number of aspects of crop protection in irrigated rice are described. The second case is crop protection in small-scale agriculture in Peru. In the Peru case the importance of the socio-economic background is emphasized.

18. Wetland rice in Sri Lanka

Rice is an important crop in the Third World. It takes up 25% of the total area given over to cereals. There has been, and still is, much research on crop protection in rice. New scientific developments are quickly transferred to the practical level.

In the past twenty years, crop protection in Asia has seen a strong shift in emphasis towards the chemical control of pests. In Sri Lanka, for example, the annual import of pesticides has increased twenty-fold between 1970-1980. The use of pesticides has been greatly stimulated by the government. Farmers could buy pesticides at a discount of 50% and there has been an intensive information campaign on the use of pesticides.

The negative agricultural effects of the high use of pesticides can be clearly seen in the cultivation of rice. For example, fields which are regularly sprayed are often troubled by the rice leaffolder and the brown planthopper, both of which seldom cause damage in unsprayed fields. These pests have only become significant with the introduction of modern, high yielding varieties which require the application of artificial fertilizers and pesticides. In Sri Lanka, however, the leaffolder seldom infests fields where pesticides are never or rarely used, even if a modern variety is cultivated. Several cases of resistance to pesticides have been noted, as well as many incidents of pesticide poisoning.

This case study deals with wetland rice in Sri Lanka but many of the aspects discussed here also apply to other parts of the world, while many of the general points may not be applicable to specific situations in Asia. After section 1.1 has provided an overview of the growth and cultivation of wetland rice, section 18.2 discusses the current crop protection methods in rice in two farming systems. Section 18.3 deals with IPM and the possibilities and limitations of IPM in Sri Lanka.

18.1 Growth and cultivation of wetland rice

The most typical feature of wetland rice is that the fields are covered with a shallow layer of water for the largest part of the growing season. This water can be supplied entirely by rainfall, provided this is adequate and spread out evenly throughout the year. Mostly, rice fields are irrigated from water reservoirs or rivers.

In Sri Lanka a traditional division of labour exists in the rice cultivation. Women transplant the rice seedlings, weed, carry the harvest to the threshing floor and prepare the harvest for storage; whilst men cultivate the soil, harvest, and thresh. On large, modern farms the first two tasks of women, transplanting and weeding, disappear. Here the rice is usually sown directly, as the women cannot manage the transplanting on their own, and it is too expensive to hire labour. Herbicides are used to control weeds and these are applied either by men or by hired labourers.

The land is ploughed, levelled and fertilized at the start of the rainy season, after which nursery seedlings are transplanted, or seed is sown directly. Adequate provisions must be made for the flow of water to and from the field, and the field must be completely free of weeds. A few days before planting or sowing, the field is flooded to drown the weed seedlings. The total duration of growth ranges from 2 to 5 months, depending on the variety of rice and the elevation of the field. The crop goes through next stages:

Seedling stage. The seedlings first of all form many roots. It is important to keep the field free of weeds at this stage. Transplanting seedlings from nurseries, the traditional method in many areas, facilitates weed control in the crop. The seedlings are big enough to compete with the weeds and it is easy to weed by hand, compared to a field where the rice has been sown directly. The advantage of direct sowing is that it saves labour, which considerably reduces costs especially on large farms. It does, however, often necessitate chemical weed control.

Tillering stage. Many leaves and new shoots are formed at this stage. Weed control must be continued until the crop covers the field, for otherwise the weeds will suppress the formation of new shoots. Thereafter, weeds become less important because they can barely establish themselves. Stem borers, such as the yellow and ink rice borer, can affect the plants at this stage, causing "dead hearts". Leaf-eating caterpillars, such as the rice leaffolder, cutworms and armyworms, can also cause damage at this stage. Sap-sucking insects, such as the brown planthopper and leafhoppers, can become a major problem. They can also transmit viruses.

Flowering stage The plants form an inflorescence. Stem borers remain a threat at this stage, causing white, dried-up flowers ("white heads") as do plant- and leafhoppers. Damage can also be caused by gall midges and various other insects, and fungi and bacteria which can enter the plants for example through wounds. The young milk-ripe grains are often sucked by rice bugs.

Ripening stage Grains which have been formed ripen at this stage. The plant dies off. Rice bugs and fungi remain dangerous, while birds and rats now also threaten the crop.

Storage The protection of the harvest is of the utmost importance, as great losses can occur during storage through various insects, fungi, rats and birds. Rice must therefore be stored under suitable conditions. This means that the crop must fully ripen in the field. The grains must be clean and dry and the seed coat must be undamaged. The store must also be clean and dry and free of any remains of previous stored products. Rice can also be treated preventative chemically before storage.

Figure 18.1. Sowing rice is a task usually carried out by women. Sowing in a flooded field prevent seeds being picked away by rats or birds.

18.2 Crop protection in Sri Lanka

18.2.1 Farming systems

Sri Lankan rice farmers can be roughly divided into three main groups on the basis of water provision. A distinction is made between: farms connected to an irrigation project where one is, in principle, assured of water (major irrigated, 36%); farms which are supplied by smaller water reservoirs, which may dry out in very dry periods (minor irrigated, 34%); and farms where the water supply depends on rainfall (rainfed, 30%). The major irrigated and rainfed farms are the most common types and this is the distinction which will be made here. The minor irrigated farms fall between these groups in most respects.

The major irrigated farms are exclusively in the dry zone in Sri Lanka and have an average size of 1.5 hectare. 8% of the farmers cultivate more than 2 ha. Nearly all major irrigated farmers take a part of their harvest to the market; 60% of them sell more than half of their harvest.

The rainfed farmers live mainly in the more densely populated, hilly districts in the wet zone. They cultivate an average of 0.6 ha, the harvest of which is mainly for their own use.

It is noticeable that many more irrigated farmers than rainfed farmers have problems with lack of funds (23% as compared to 4%). The costs of input and labour are, of course, much higher for them. They do have funds when they have sold a part of their harvest and this is not the case with the rainfed subsistence farmers. Rainfed farmers often have a part-time job in addition to their farming activities, a part of their income from this being presumably spent on the cultivation of rice. A subsistence crop is very valuable, as it is relatively expensive for farmers to buy rice. The market price is about three times as high as the price a farmer receives for his rice and when this price is compared to production costs, the comparison is even more unfavourable.

Nearly all the farmers plant modern high yielding varieties of rice, 42% of them sowing resistant varieties. These are mainly farmers in the dry zone, which was seriously affected by the brown planthopper in the past. These farmers are, however, not always aware of the fact that the variety they use is resistant, and often choose a variety for its higher yield.

A few rainfed farms (3%) still plant the traditional varieties which are also very resistant to local pests and diseases, but have a much lower yield. These varieties are often chosen for their good flavour.

More rainfed than irrigated farmers (22% as opposed to 8%) still use traditional methods of controlling pests such as sprinkling ashes, using sticky parts or repellant extracts of plants, and singing mantras. However, this is often done in conjunction with the use of pesticides.

A good example of a traditional method with a scientifically proven effect is the planting of African marigolds on the dikes along the rice fields. Marigolds seem to have a repellant effect on various insect pests. Some farmers sow these plants next to their fields in the belief that their crops will be blessed if they offer the flowers in the temples, most of the farmers being Buddhist or Hindu.

About half the farmers prepare pesticide solutions incorrectly. On the whole, the farmers are aware that pesticides can have a bad effect on the health, especially of those who apply them, but also for everyone if residues remain in the food. There is less awareness about the negative effects of pesticides on the environment, and the effects in the long term are largely overlooked.

Irrigated farmers are better informed than rainfed farmers about which pesticide is to be used against a particular pest. In general, the wrong pesticide is often used. A quarter of the farmers get advice from the extension officer on the pesticide they should use, and 8% from the supplier of the pesticides.

The majority of the farmers (61%) use too low a concentration of the insecticide with the result that the pest is inadequately controlled. Most of the labels on pesticide containers give instructions in the local languages (Sinhalese, Tamil and English), but not all farmers can read. One reason often given for using too low a concentration is lack of money.

18.2.2 Insect pests

The farmers consider the brown planthopper, the rice leaffolder and stem borers to be the most serious insect pests. Irrigated farmers in the dry zone have often more trouble with pests, especially the brown planthopper, than rainfed farmers.

Figure 18.2. If rice is transplanted, this should be done in rows, so that weeds are easier to hoe.

137

On the average, a crop is sprayed with insecticides 1.7 times, and in this respect there is little difference between the two farming systems. Compared to rice farmers in the Philippines, where average frequencies of 2.5 sprayings per crop were recorded, the use of insecticides in Sri Lanka is not high.

The greatest difference in the average number of times a crop is sprayed can be seen when the farms are considered according to districts. Probably the extension services, which are organized by districts, influence the farmers' use of insecticides. Another important factor is the occurrence of epidemics of certain pests in the past. In areas where this is the case more pesticides are used.

The insecticides most commonly used in Sri Lanka are monocrotophos (mainly against leaf-eating caterpillars), BMC (a selective insecticide against plant- and leafhoppers) and metha-midaphos (not recommended for rice cultivation by the Ministry of Agriculture).

18.2.3 Weeds

The control of weeds, especially grasses, is seen as the most serious problem by most farmers. This could be because in crop protection extension services the emphasis is often placed on the chemical control of insects. Irrigated farmers have more trouble with grasses than rainfed farmers.

The most commonly used herbicides are MCPA (against sedges and broad-leaved weeds, although in 40% of the cases it is also used against grasses) and 3,4-DPA (against grasses and sedges). These herbicides are used by 70% of the irrigated farmers and by only 34% of the rainfed farmers, and then usually only once. The difference between the two farming systems can be attributed to the fact that irrigated farmers more often sow directly in the field, which makes weeding by hand difficult.

18.2.4 Fungi and bacteria

Fungicides for the control of fungi are hardly used and fungal infection is often not even recognized as such. Farmers sometimes spray insecticides against fungal and bacterial diseases.

18.3 Integrated pest management in rice

18.3.1 The FAO programme

As mentioned earlier the Food and Agriculture Organization initiated an IPM programme for rice in South and South-east Asia in 1980. In collaboration with the agricultural extension services and research institutes of the countries concerned, the technology developed by IRRI was adapted to local conditions. The programme developed for Sri Lanka is presented here.

The aim of the programme is on the one hand to train rice farmers in IPM methods and also to ensure that the acquired knowledge is applied on the farms. On the other hand it aims to create a general awareness amongst the farming population of the negative effects of chemical control. In Sri Lanka a multi-media campaign has been developed for this purpose, in which, for example, radio announcements are used to create resistance to persuasive radio advertisements for pesticides.

The IPM training programme is embedded in the structure of the government's agricultural extension services. This service operates through the so-called 'Training and Visit System', an extension strategy developed by the World Bank. The extension officer at village level, the KVS

Figure 18.3. During the tillering stage, it is very important that no weeds inhibit the rice plant forming new stems, otherwise less panicles will be formed.

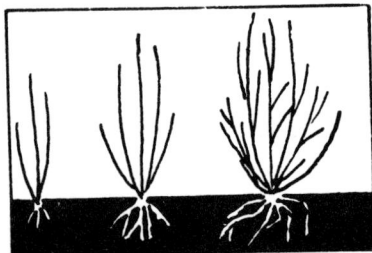

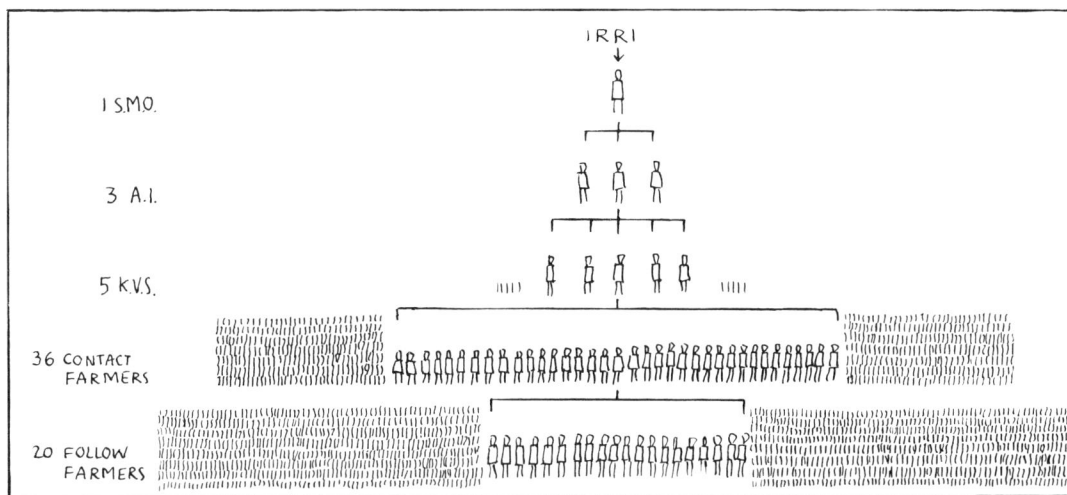

Figure 18.6: How to pass on the IPM-message to millions of farmers in Sri Lanka? Each stripe signifies one person to be trained.

Figure 18.4. All kinds of pests attack the rice plant during growing: fungal diseases, aphids sucking from the grains being filled, stem borers, planthoppers, and so on.

(Krush Vyaphthi Sevaka) has 36 contact farmers under his or her care. These contact farmers undertake group training once every two weeks and should also be visited at home by the KVS. The IPM technology has become a part of the two-weekly training programme, which covers all aspects of rice cultivation. Each contact farmer is responsible for passing the acquired information on to 24 farmers.

Some five KVSs are trained and guided by an Agricultural Instructor (AI) who, in turn, is trained by others, including an expert on crop protection (SMO – Subject Matter Officer). All SMOs are intensively trained in IPM through the FAO programme, as are a selected group of AIs and the KVSs, thus maintaining direct contact with the people working at village level. One advantage of this extension system, when it works well, is that many farmers are reached quickly. However, the danger of superficiality in the transfer of information through so many levels is great. A number of questions arise here: Which farmers are ultimately going to be reached? How is the KVS going to choose the contact farmers (modern farmers who can quickly show results, thus increasing the KVS's chances of promotion? economic interests? favouritism?)? How does a contact farmer select the 24 farmers? Will the contact farmers pass on the information correctly (will they seek to strengthen their own position? didactic aspects?)?

To eliminate these problems to some extent, the FAO organizes, in addition to the regular two-weekly training programmes for contact farmers, training programmes lasting a full season. KVSs give specific training in IPM, to a fixed group of 20 - 25 farmers who have applied for this training, once a week during the growing season. A much greater amount of information can be passed on in these training courses.

In the last two years (1984-1986) the FAO project in Sri Lanka has extended to 19 of the total of 26 districts. In 1986, 17,000 farmers could be trained in each season.

The training programmes for IPM cover an extensive set of recommendations, the most important of which, as applicable to rice cultivation, is discussed in the frametext. For further, general methods of IPM, see Part 2 of this book.

Figure 18.5. Any damage done to the product to be harvested will cause serious losses. Birds and rats are main attackers of ripe rice, and some varieties are susceptible to shedding. The grains may be infested bij storage pests at this stage.

IPM measures in rice

Weed control. Before the crop has grown enough to cover the whole surface of the field, it should be kept free of weeds. This can preferably be done mechanically, either by hand, or with the help of a tooth or roll weeder. These weeders are, however, not yet universally available and can only be used if the rice is planted or sown in rows.

Fertilization. Adequate fertilization is a pre-condition for a strong crop which can tolerate a certain degree of infestation. Over-fertilizing with nitrogen leads, however, to an increase in infestation particularly by the brown planthopper and the stalk borer.

Resistant varieties. The use of resistant rice varieties is one of the most effective components of IPM. In Sri Lanka only three highly resistant varieties are available:

one resistant to the brown planthopper and two resistant to the rice gall midge. In addition to resistant varieties, fast growing varieties are also recommended. On most fields in Sri Lanka two rice crops are grown per year and the use of early varieties limits the number of generations of insects.

Synchronous planting. Groups of farmers with adjoining fields are encouraged to sow varieties with the same growing period at about the same time so that all the fields are cultivated at the same time. A field that is too early or too late gives insect pests the opportunity to survive

Protecting and stimulating natural enemies. Rice fields where pesticides have never been used house a rich variety of the natural enemies of pests. By using as little pesticides as possible, as many

natural enemies as possible can be saved. Granulates will destroy relatively fewer natural enemies than insecticides which are applied to the leaf surface. Planting pulses on the dikes along the rice fields stimulates the presence of various parasite wasps of rice pests as they find nectar in the flowers. Providing nests for insect-eating birds can also contribute to biological control.

Chemical control. It is important that the farmers use a proper pesticide for a pest, apply it in the correct formulation and concentration, and at the right time. It is desireable to use selective pesticides in order to preserve the natural enemies. The problem is, however, that such products are often more expensive and not available everywhere.

18.3.2 Monitoring and damage thresholds.

To determine the extent of pest populations before taking control measures, the rice plants must be sampled carefully once a week. Most farmers visit their rice field every day. Although they seldom take careful samples, as required in IPM, this habit provides a good starting point for their training.

It is very important to be able to identify the pests as well as the natural enemies of pests. The numbers of harmful insects or the symptoms of damage are related to fixed damage thresholds. In Sri Lanka, for example, this value has been fixed, in the case of the brown planthopper, at 8 hoppers per hill (bundle of rice plants). The damage threshold varies during the course of the growing season, depending on the vulnerability of the rice plant in its various stages of growth. The values used at the start of the FAO programme were taken from Kerala,

India, and soon proved to be too low for the Sinhalese situation. They were then adjusted on the basis of experiments made on farmers' demonstration fields, and led to a reduced use of pesticides.

To be able to take the right decision on whether it is necessary to apply a pesticide or not, the farmers are taught to take some factors into account. In addition to factors such as the numbers of insects or the degree of damage related to damage thresholds, the stage and condition of the crop, farm conditions such as the water situation, weather conditions and the number and importance of natural enemies should be considered.

On the whole, the sampling procedure and the use of damage thresholds is found to be complicated. It is important that the farmers not only acquire this knowledge, but also regularly apply it in their farming practice. Social reinforcement is important to trust their own judgement.

18.3.3 Suitability of IPM

Using the techniques of IPM, as disseminated by the FAO, should lead to higher profits for rice farmers, either through savings on pesticides or through higher yields as a result of more effective pest management. The first factor is of particular relevance to the major irrigated rice farmers, whilst the second factor would be more relevant to the small-scale farmers.

The whole range of recommendations seems to make a rather great demand on the farmers as far as the learning and use of new methods is concerned. However, many of the recommended practices are already used traditionally by the farmers, such as the thorough cultivation of the soil. The farmers simply have to become aware of the value of such practices in the context of crop protection.

A number of the traditional methods, forgotten by scientists, have been taken up again and included in the list of recommendations. Examples of this are the use of sticky parts of plants, such as those of papaya trees to catch insects, or dragging a banana stem over the young rice plants to control thrips, caterpillars and sedges.

The most important aspect of IPM, the weekly sampling of the crop, will be especially time consuming for the major irrigated farmer. For the rainfed farmer, the quality of his visit to the rice fields must be changed.

The IPM approach should, in theory, not cause any additional costs, but money must be available at the right time to buy the required inputs. This could prove to be a problem particularly for the major irrigated farmer. A further condition is that the required inputs, such as resistant varieties and selective pesticides, should be available locally and at the right time. This is often not the case for rainfed farmers, who live in the more distant, inaccessible mountain regions. Farmers who are difficult to reach are also visited less often by extension officers, and have greater difficulty in attending training sessions.

On the whole it seems that IPM in rice could be successful both on irrigated and rainfed farming systems if all the farmers were to have good access to a thorough training. Cost savings can in any case be achieved, as illustrated by the fact that on the demonstration fields of irrigated farmers the number of sprayings decreased from seven to one per season. Whether the farmers can adopt the recommended methods so thoroughly that IPM will work on the long term, remains to be seen.

19. Small-scale farmers in the Andes

Usually pesticides are not much used by small-scale farmers in the tropics. When pesticides are introduced this is often to obtain increased production for the market. In traditional agricultural systems, production is primarily intended for household consumption. In this system of production a low yield is not a great problem, as long as it is stable yearly. A low crop yield as a result of pests is not very significant. In some cases pests occur very clearly, the farmers do not see these as a problem.

The situation in Peru is no exception. In this case study we provide an illustration of the problems caused by pests and pesticides for farmers who are in the transition phase from subsistence farming to a market oriented system. We will discuss the situation in the department Ancash, an area in the Andes in the north of Peru.

19.1 The natural environment

The entire Andes range is typified by steep slopes. There can be a difference in altitude of more than 1500 metres between areas which are no further than 10 km apart from each other. These differences in altitude create large climatic differences within relatively small areas. As a result, different crops can be cultivated at different altitudes, according to their optimal climate.
In figure 19.2-6 we show that at any given altitude at least one pulse (bean, pea, lentil, broad-bean or tarhui) and one tuber crop (potato, oca, olluco or mashua) can be cultivated. Up to an altitude of 4000 metres there is also at least one cereal crop suitable for each altitude (maize, wheat, barley or quinua). A farming family with plots at different altitudes thus can produce a very varied range of food crops. That is why farmers in the Andes have been aiming at a so-called "vertical orientation" of their farm; for example, they would rather have three plots of 1 hectare each at 2800, 3000 and 3500 metres than a plot of 3 hectares at one altitude.

19.2 Traditional agricultural systems

In this rugged environment of steep slopes and local differences of climate, three kinds of farm have developed from the 16th century to the middle of this century: large ranches, the (marginal) subsistence farms of farm labourers, and farmers organized into agricultural cooperatives.

Large ranches. Up to about 1970, more than half of the department of Ancash was in the possession of large landholders who mainly kept cattle and sheep. These ranches ranged in size from 50 to 10,000 hectares. The more fertile areas of land were irrigated and used to produce lucerne as cattle fodder or sometimes to produce food crops such as potatoes, maize or wheat. From the 50s onwards the farmers most often control pests in their livestock or crops with chemicals. There was a purely physical reason for the emphasis on the production of meat and wool by the large ranches. Up to 1950 the donkey was the only means of transport between the Andes range and the rest of Peru. The distance to the nearest road or harbour was at least 200 km. In such a situation, livestock was the only agricultural product suitable for inter-regional trade. Cereals, pulses and tuber crops were unsuitable for a production system which was aimed at generating capital in a naturally poor area.

Farm labourers. The ranch owners employed about one-fifth of the local population as labourers. The labourers were usually paid in kind, and were allocated a piece of marginal land by their landlord, on which the family could grow food crops for own use.

Farmers' cooperatives. According to the 1981 census, a good half of the farms in Ancash were members of a cooperative, a ***communidad campesina***. Such cooperatives are still found in the entire Andes region, not only in Peru, but also in Bolivia and Ecuador. In addition to Ancash, there are seven other departments of Peru which are situated partly or entirely in the mountains. In these departments 46 to 65% of the rural inhabitants are members of a cooperative. In these cooperatives no farmer owns private land; the land belongs to the cooperative. This is regulated by law. In principle, all the farms in a cooperative are of the same size; in practice there are some differences in size, as a result of, for example, marriage or death. Land cannot be bought or sold by individuals, but a large part of the land suitable for agriculture is given in usufruct to the families in the communidad. The farmer and his family is therefore both the production and the consumption unit.

In the context of plant protection, a number of farm cooperatives take binding decisions on issues such as crop rotation, sowing and harvesting date at their members' meetings. In such cases, therefore, important decisions on crops and crop protection are not taken by the family. Such arrangements help the cooperatives to achieve a considerable reduction in pest populations. In contrast, infection or infestation from a neighbouring field is common in areas where each farmer decides on his or her own crop rotation.

The way in which cooperatives are organized makes it a lot easier for each farm to have plots of land at different altitudes. An individual farmer with plots at some distance from each other, has difficulty in controlling his fields adequately. A cooperative can combine a number of tasks such as the construction and maintenance of irrigation canals, protection against theft before harvest, and the transport of fertilizers and the harvested crops. In addition to this, many cooperatives have their own systems of social security, such as neighbour help and reciprocal labour. Especially in the case of the illness or death of male members of the family, neighbours will help in exchange for a part of the harvest. The loan of draught animals also falls under the rules of reciprocal help, with the result that the available draught animals are used more intensively than in villages where the land is owned privately.

19.3 Developments in the agricultural systems

Up to the 60s, agriculture in Ancash had barely changed. Since then, mainly government regulations have caused a number of changes.

Expansion of the road system. Since 1960 several new roads have been build between the

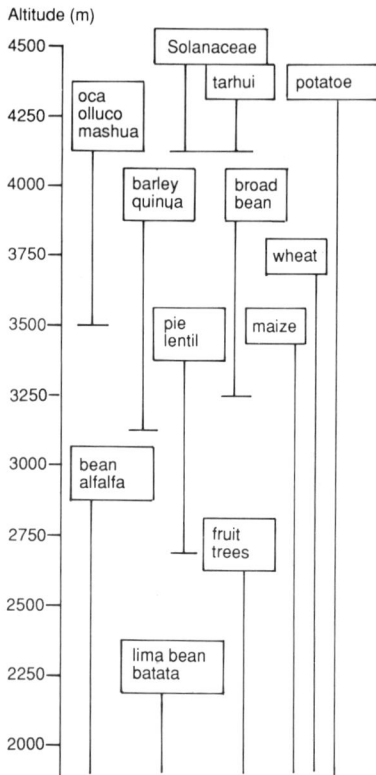

Figure 19.1. In the Andes various traditional crops can be found which are almost exclusively eaten by the Indian people. Examples of these are *tarhui*, a pulse, the tuber plants *oca*, *olluco* and *mashua* and the protein-rich *quina*.

Andes mountain and the rest of Peru. This opened up prospect for the transport and sale of agricultural products by lorries and also for the transport of artificial fertilizers and pesticides. Many ranches then changed over to use their irrigated fields for the permanent cultivation of maize, wheat, potatoes and fruit, destined for the city consumer.

For the farm labourers these changes brought little improvement. The transition from keeping livestock to permanent cultivation on the ranches demanded a much larger labour input. The payment in kind remained low and their own marginal subsistence farms were too small to produce surpluses for the market.

The cooperatives could use the improved transport to market their surpluses, but they were at a great disadvantage compared to the ranches as a result of the isolation of the farms high in the mountains, far from roads. The hiring of pack animals was a considerable expense, resulting in lower incomes compared to the ranches.

Land reform. In 1968, the government announced land reforms in the whole of Peru. Under the motto of "the land belongs to those who work on it" the large land holdings were to be confiscated and divided amongst the labourers and farmers in small farms. On account of the distances in Peru the implementation of these reforms took many years. Especially in the isolated mountain regions, the large land owners had the opportunity to sell off their land in farms of about 10 ha before it was confiscated. This capital was often invested in property or the trade in lorries. The current landowners are often traders and officials, who employ labourers to work on their land, the same labourers who were previously in the employ of the landlords. The new, medium sized farms produce on a commercial basis. They invest in fertilizers, pesticides and improved seed and planting material. They receive intensive guidance from the agricultural extension services, while agricultural research carried out at universities and by the Ministry of Agriculture is also mainly geared towards this kind of farm.

Most of the cooperatives have gained few direct benefits from the land reforms. A few have been able to expand their cultivated land with the land of former large land holders. These are often higher areas which are not irrigated and which are subject to night frost. More than one crop per year is still not possible for most of the cooperatives on account of the climate and the limited capacity of their irrigation canals. Furthermore, it is not cost effective to use artificial fertilizers, pesticides and improved seed or planting material because the harvest is quite uncertain.

Figure 19.2. Potatoes grow very well on the high cold fields in the Andes, where they stay relatively clean from pests.

19.4 Changes in agricultural policy

Since 1980 the agricultural policy of the government of Peru has been geared more and more to meeting the demand for cheap food by the rapidly growing and increasingly pauperized urban population, and hence on replacing food imports. For the farms in the mountainous areas this means that only those crops are encouraged which have been imported in large quantities to date, or which are much in demand by the urban consumer. This is mainly maize, potatoes, wheat and barley. Credit facilities and minimum prices apply only to these crops. This crop-bound stimulation policy has broken the traditional cropping pattern, of which a result is the increase of pests.

Farmers who get credit do not get this in the form of cash in hand, as experience has indicated that this leads to unwise spending on alcohol etc. Credit is only extended in the form of cheques, which can be exchanged for artificial fertilizers, pesticides and seed. There is no quality control on these items and the seed and planting material introduced have been repeatedly found to be infected. The government has furthermore announced a number of changes in the law, aimed at dismantling the cooperatives. According to the view of the present agricultural policy "modern agriculture" cannot co-exist with traditional structures. Since 1982 farmers' cooperatives can be dissolved by a majority vote of the members. The resulting privatized farms are then eligible for agricultural credit, with the land as surety.

19.5 Crop protection

The above mentioned developments have had various consequences, also for the importance of pests in crops. The new middle-sized farms permanently cultivate maize, potatoes and wheat, and therefore pests occur much more frequently than before and are controlled chemically. This category of farm has easy access to extension on crop protection. Small-scale farmers and cooperatives are in a very different position. They cannot afford to ignore the shift from subsistence farming to partially market-oriented farming. They are, in fact, highly dependent on the income from the sale of produce to meet their subsistence costs. Above all, the pressure on cooperatives to integrate in the national economic structure has had destabilizing economic, social and ecological effects, so that pests become important loss factors. The following examples illustrate how the interests of the planners of agricultural research and extension do not coincide with the interests of the disadvantaged farmers.

19.6 The reform of potato production

19.6.1 Traditional potato farming.

The Andes area is the home of the potato. Some 10,000 years ago the first potato races were domesticated here. The local potato races, of which there were several thousands, have been cultivated in small fields at high and medium altitudes for a long time, mainly for subsistence. Each variety had its own cultural characteristics and culinary use. The farmers retained their own seed potatoes for the following season. Seed potatoes were constantly exchanged and traded between families so that each family could choose the races that best suited their domestic needs, and which would furthermore give them some assurance of a reasonable yield. There was mixed planting of several varieties in each field, for this reason these varieties were called Huyaco varieties ("huyaco" means mixed).

Long interval crop rotation and the use of animal manure in this system ensured that soil diseases caused little damage. Risks for pests were spread by the large genetic variety within each potato field. However, at medium and low altitudes the quality of the seed potatoes did gradually deteriorate. After each season they became more and more infected with virus diseases, so there had to be a constant supply of relatively healthy seed potatoes from the high cold zones where the virus transmitting insects (mainly aphids) have little chance of survival. Moreover, the Peruvian farmers have from time immemorial used the system of producing healthy planting material from true seeds in the highest areas. Neither pesticides nor artificial fertilizers were used in potato production.

Figure 19.3. Pest-free Huyaco seed potatoes are taken down to fields at the middle altitudes.

19.6.2 The "Blanca" system.

The last twenty years the government is introducing a new system of potato production in the Andes. This system is mainly geared at the urban market. Farmers use entirely different varieties not based on the system of producing improved varieties in the high zone. The farmers are motivated to produce for the market by their increased need for money. Farmers now also want to buy consumer goods and agricultural inputs. The increased production for the market had led to more than a doubling of the area planted with potatoes.

What motivated the authorities to introduce a new system of potato production? In the fifties and sixties a technical solution to food deficits was aimed, also in Peru. The authorities thus stimulated a project which aimed at providing all potato producers in Peru with "certified" seed potatoes of high quality, linked with the use of fertilizers and pesticides.

Further research by the International Institute for Research on Potatoes in Peru (CIP) at the end of the seventies indicated that these ideas on the lack of correct cultural methods, technology and uninfected seed potatoes did in fact simply arise from prevailing prejudices and did not apply to potato production in Peru. Nevertheless, the project continued.

The project at first had the most influence in areas where there was potential for the increase of commercial potato production; that is, along the coast and in the wide, easily accessible valleys of the Andes. In the areas above 3500 metres, mainly used for subsistence farming, the improved potato Blanca was rarely found. The seed potatoes for this variety must be bought from PRODERM, a project sponsored by various national and international organizations. The seed potatoes that the farmers keep themselves degenerate within 3 to 4 years and must then be replaced. The Blanca varieties are not mixed in the fields as this creates problems for marketing. Pesticides and fertilizers are used. Under these conditions Blanca gives a higher yield than Huyaco varieties. However, they do not store well and are inferior in taste (bitter and watery).

In recent years a few Huyaco varieties have been adopted by PRODERM, which multiplies them centrally and then distributes them. These races, called **Nativa**, do not have the culinary and storage drawbacks of the Blanca and are therefore more suitable for use by farm families, while any possible surpluses can be marketed. In fact, they fetch better prices than Blanca. From the agricultural point of view they are also superior as they are better suited to local conditions than Blanca. For the rest, however, cultivating Nativa is the same as cultivating Blanca. Nativa functions as a bridge between the "old" and the "new" agricultural systems. In this way "modern technology" is penetrating the domain of local potatoes. However, the farmers are not satisfied with a few Nativa varieties. As already mentioned, each of the main Huyaco varieties has a special place in the kitchen and often forms a welcome change in an often meagre diet.

19.6.3 The new pests of Huyaco

In the middle zones the increase in the production of the Blanca and Nativa varieties for the market appear to have environmental effects which are disastrous for the production of Huyaco. In the first place, it is uneconomical for the farmer to plant Huyaco on the same fields as Blanca and Nativa. Huyaco plants are often infected with viruses to which they themselves are fairly resistant. The transfer of these viruses to Blanca plants causes large-scale damage. But the production of Huyaco also suffers from the presence of Blanca and Nativa. Before these varieties were produced such pests as *gusano* and *gorgoju de los Andes* (**Premnotrypes spp.** and **Bothynus sp.**) were either not found or were not a problem, the farmers think that

Figure 19.4. Clean Huyaco-seed-potatoes from higher altitudes are planted at middle altitudes.

147

Toiling on cold, distant fields for a good yield below

In the high areas the temperatures are low. The long fallow period allows the soil structure to recover and suppresses the organisms which cause the potato diseases. Cattle grazes in the fallow fields and so fertilize them naturally. The land in the high zone (higher than 3500) is ideal for potato cultivation. The highest fields have always produced disease free seed potatoes, as most of the diseases that affect potatoes do not flourish at that altitude. This is why seed potatoes from these areas can be used much longer than those from the lower-lying areas: from 8 to 10 years. Therefore there is a regular exchange of planting material between fields at different altitudes. The use of pesticides in the high zones is limited to actual outbreaks of disease. Artificial fertilizers are also seldom used here. Mechanization is virtually impossible. The transportation of potatoes, artificial fertilizers, pesticides and equipment to the higher zones is very costly and this is why market-oriented production develops very slowly here.

In the middle zones pesticides are used fairly generally. Most farmers use the knapsack sprayer, occasionally granules, mainly dieldrin. Mechanization is possible on the flat valley lands. The use of costly inputs such as seed potatoes, artificial fertilizers and pesticides here ensures a somewhat increased yield. In addition, the milder climate reduces the chance of crop failure through frost or hail, and the subsequent loss of these costly inputs.

The technical possibilities in the low zone are the same as those in the middle zone. The use of artificial fertilizers and pesticides is common. There is much mechanization. As this zone is much warmer, only a limited area is planted with early varieties, which do not keep well and are only suitable for immediate sale or consumption.

Figure 19.5. Seed potatoes from high altitudes give a good yield at lower altitudes the first few years.

the expansion of these pests was encouraged by a greater use of pesticides (maybe via the reduction of the number of natural enemies). There are indications, that not only the Huyaco, but also the Nativa varieties are disappearing from places where Blanca is introduced. The new pests affect Nativa more seriously than the Blanca varieties. The increase in the cultivation of potatoes in general can also well have led to this increase in pests.

In the high zone, however, the area planted with Huyaco is reduced for other reasons. The soils of the high zone require great care, and the potato requires a very good soil structure. Farmers preferably use the manure of their cattle on the potato fields. Application of animal manure improves the soil structure. The farmers say that artificial fertilizers impoverish the soil and make it hard. If they do use artificial fertilizers they always mix this with animal manure. Previously soil diseases disappeared during the long fallow periods and the cattle improved the soil fertility. This made the Huyaco varieties fit in perfectly with the farming system that was most suited for the vulnerable ecological system of the highest agricultural areas in the Andes. Possibly as the result of the land reforms at the end of the sixties, small-scale farmers now have more land than before, thus they need more seed potatoes. The fallow period in the high zone is therefore reduced to a half or a third, giving rise to reduced potato production and increased infection by pests such as nematodes, fungi and viruses, and thus infected seed potatoes, while the high zone should provide healthy Huyaco seed potatoes for the lower zones.

The quality of the Huyaco seed potatoes thus remains a problem. It does not seem as if the process of intensification will continue in the high zone as an increasing number of farmers now no longer make use of their land rights in that zone, and the other farmers do not have the seed potatoes and the labour needed to use the land that has come free. It also frequently happens that the cooperatives neglect the highest fields first.

Without healthy Huyaco-seed potatoes it is not possible to produce this variety in the lower zones. The farmers confirm that the production of Huyaco in the higher zones is a precondition for its continued existence. As much as they regret the disappearance of the Huyaco varieties, this is not a sufficient reason for the farmers to continue to work in the cold and distant fields. For each individual household there are better alternatives for the use of labour. In this way many varieties have already irrevocably disappeared.

19.6.4 Potato crop protection

The use of pesticides has increased in the wake of the new seed potatoes, while the problems of plant protection have changed, but have not been reduced. The PRODERM project has made its own theories have become fact. Where, until recently, Huyaco had surpassed Blanca in its resistance to pests through its diversity and flexibility, the position has now been reversed, and this change cannot be stopped.

The above example has illustrated how close is the relationship between method of cultivation and the plant protection problems that can be expected. It is clear that the use of pesticides and neglect of the principles of crop rotation have played an important part in the loss of the Huyaco varieties. Even if we accept this loss, it is necessary to change to a more sensible form of plant protection. At present dieldrin and other pesticides prohibited in developed countries form a large part of the pesticides used in potato production in Peru, although recently a campaign has started to forbid the use of thirteen dangerous ones. In the area described above, many farmers even broadcast these pesticides by hand!

Figure 19.6. After some seasons the potatoes get infected more and more, and yields drop.

A second example, taken from maize production, illustrates that there are possibilities for a different approach.

19.7 The control of corn earworm in maize

From about 1970 the corn earworm **Heliothis zea** is the most significant maize pest in the mountainous part of Ancash. In the United States it is known as an insect pest in cotton and maize. It has been estimated, on the basis of field experiments made on farms in 1979 and 1989 that up to 65% of the maize crop in Ancash is affected by this caterpillar.

19.7.1 Life cycle of the corn earworm.

How does a corn earworm look? The adults insects are moths. The females lay their eggs on the young pistils of the maize flower. The eggs hatch into larvae which eat their way down along the pistils and inside the cobs. This phase lasts six to eight days. Once the larvae have reached the developing kernels the real damage starts. In the following three weeks they grow from a size of 6 mm to 8 cm. One larva can eat up to a third of all the kernels on a cob. After

149

this growth stage they eat their way to the outside, right through the cobs, drop to the ground and dig themselves into the soil to pupate. The pupal stage lasts from two to nine months. The larvae leave a hole of about 4 mm in the cobs, through which rainwater comes in and smaller insects which lay their eggs. This causes secondary damage through fly larvae and fungi which also make the remaining kernels useless.

19.7.2 Difficulties to control the corn earworm

The biggest problem in the control of corn earworm is the short time during which the eggs and larvae are on the surface of the cob and can thus be reached by a pesticide. Once the larvae have eaten their way inside the cobs there is no point in spraying with pesticides. The usual control method advised by the extension services consists of repeated spraying of the pistils at intervals of three days. According to this spraying advice, the spraying must start from the time when the pistils become visible until the last pistils have dried up. Adherence to this advice results in an average of six sprayings per field with a total of about 16 kg of Sevin (85% carbaryl) or the organosphosphorous compound Dipterex (80% trichlorphon) per hectare. For farmers with non-irrigated maize at zones higher than 2200 metres implementing of this spraying advice is barely possible. A large part of the maize production is intended for own consumption and this makes expenditure on pesticides very cost ineffective. For most of these farmers a knapsack sprayer alone is already beyond their means. In an area where it is advocated to spray every three days it is also nearly impossible to hire a sprayer.

Furthermore, it is especially the non-irrigated farms which suffer the greatest damage from the corn earworm. There usually genetically heterogeneous local varieties are sown. This means that the period during which the maize is susceptible is longer than with hybrid seeds which produce uniform plants developing at the same rate. Although the local varieties are in themselves not more susceptible to this pest, the chance of infestation is greater and the execution of a spraying programme is more expensive.

The corn earworm has spread not only in Ancash, but also in the remaining northern departments of Peru. In the south the corn earworm causes little damage. It has proved to be impossible to find a solution to this kind of problem within the structure of the agricultural university and the Peruvian Ministry of Agriculture. The top officials who determine the policy of both these institutions have a strongly centralized approach. The agricultural extension service also shows a preference for uniform research topics at national level and, linked to this, for uniform training courses and printed matter. A request to create decentralized research into pests, such as the corn earworm, was rejected on the grounds that this would lead to amateurism and chaos which could not be controlled from the capital.

19.7.3 A regional problem-solution approach.

For this reason a few individual researchers set up a research project on the control of the corn earworm, together with students and, above all, farmers from the cooperatives. They researched the following aspects: the potential for biological control with the help of parasitic wasps of the species *Trichogamma*; the potential for killing the pupae by intensive tillage; and the injection of a pesticide into the maize cobs.

The final results of this research is that the biological control of corn earworm with *Trichogamma* is not effective in practice, as these wasps do not reproduce under field conditions. Killing

Figure 19.7. In order to get good yields at middle altitudes, clean Huyaco seed potatoes have to be cultivated at high altitudes.

Figure 19.8. An extension ("capacitacion") session in the Andes.

the pupae by repeated tillage also had no effect as the adult moths move quite a lot and quite far. This means that there is a constant supply of new moths which fly in from elsewhere to lay their eggs. This research project did show that the Heliothis problem has increased in the last few years, because irrigated cultivation is now carried out throughout the whole year, so that the build-up of the pest population is not interrupted during a cropless season, as previously.

The most practical way of limiting the damage caused by the corn earworm proved to be chemical control by means of injection. The practical advice of the farmers is as follows: from the time that the first pistils appear in the field ten plants should be examined every day for the presence of eggs. These are about 1 mm in diameter and therefore visible to the naked eye. When the first eggs have been sighted, preparations must be made to inject within eight days. In this period there should be daily control on whether the number of eggs increases or not. If there are doubts about the extent of the infestation, fifty plants are sampled. This means that the tops of the cobs are carefully opened and checked to see whether the small larvae (of about 6 mm) have penetrated. If this is the case in more than ten cobs, then the action threshold has been reached and the infestation is sufficiently severe to warrant control. The injection method works as follows: the carbamate Sevin or the organophosphorous compound Dipterex is mixed into a solution ten times as concentrated as for normal spraying. It is preferable to use carbamates rather than organophosphorous compounds, as the latter are far more

151

Figure 19.9. Treating an infested maize crop.

poisonous. Carbamates also leave hardly any toxic residues in the harvested product. This solution is injected into every corn cob with a disposable syringe fitted with a thick needle. Veterinary syringes and needles are ideal for this purpose. The dose of 1 cc per cob must be strictly followed. More than this amount leads to rotting through too much moisture inside the cobs. Less that 1 cc per cob or a more diluted solution does not kill enough larvae. The experiments made to compare concentrations and dosages were made on the fields of the farmers, and they were very enthusiastic about the application of this method.

The most important advantage of the injection method is that a knapsack sprayer is no longer necessary and that much less pesticide is used; an average of 2,5 kg carbamates per hectare, as compared to the 16 kg required by the method recommended by the agricultural extension services. In addition to this, the method causes less harm to the environment through a better targeted use of the pesticide and a reduced rate of use.

It is amazing to see how the farmers use the new sampling method, although this is a "new" skill for them. Most important is the recognition of the connection between eggs on the pistils and a half-eaten cob three months later at harvest time. The injection method is, of course, labour intensive: an estimated 18 working days per hectare. However, this is light labour, so women and even children can help. The injection method however is not cost effective for farms which hire labour.

19.7.4 Acceptation of the new method

Four years after these recommendations were first made, the injection method is used on an estimated 30% of the farms which produce maize. This number is steadily increasing through the exchange of information between farmers. The agricultural extension service has since then also incorporated the method in its training programmes. The extension officers are not very enthusiastic about the method, however, and do little to spread it. This attitude is mainly caused by their business connections with the suppliers of pesticides in the departmental capital, for whom they function as representatives. When one considers the salary earned by an extension officer, it is not surprising that they indulge in this form of small-scale corruption. However, their attitude to the farmers is once again evidence of the fact that they cannot be trusted, as they represent interests other than those of the small-scale farmer.

19.8 Conclusions

Agriculture in the Ancash region of the Andes mountains has undergone changes only in the last thirty years. The farmer's co-operatives were well adapted, as a form of business, to the environmental situation; even the large estates, normally objectionable on the grounds of social justice, operated with a certain logic within the technical limitations and natural possibilities available in the area.

Now, owing to improved transportation, land reforms and political regulations concerning agriculture, the farmer's co-operatives, i.e. more than half the rural population in hilly areas, are involved in the change-over from self-sufficiency to market-oriented production. Through its agricultural policy and a system of credits and minimum prices, the Peruvian government encourages the cultivation of merely a few types of crops, intending in this way to bring the farmers' co-operatives into the national market for food products. The result is that the co-operatives are forced into an unstable situation, economically, socially and ecologically: their competitive position is weak against the medium-size commercial undertakings; dissolution of the co-operatives diminishes their own social supply system and the unbalanced stimulation of just a few of the crops brings traditional crop rotation and mixed cultivation into disuse.

The developments in potato cultivation indicate this problem clearly. The original potato varieties were superseded by improved varieties, a trend which not only brought the inevitable problems of crop protection and pest control, but which also severely disturbed the traditional agricultural system, particularly among the farmers' co-operatives in the mountains. The arrival of new varieties worsened the position of the small potato farmer in comparison with his bigger colleagues, rather than improve it: they do not have the money to purchase the necessary inputs, there are transport problems, limited possibilities for mechanization and, last but not least, the social situation is unbalanced. This kind of problem is not limited to the cultivation of potatoes; farmers are faced with the same kind of situation in the case of other crops. Presently, pests are becoming an ever greater damage factor in crops. Advice for control, as prescribed by agricultural research institutes, extension service and dealers in pesticides is not geared to the situation of poor farmers. The prescribed quantity and frequency of spraying is far too high, making it unremunerative for small farmers and farmers' co-operatives. Research to appropriate pest control against the corn earworm proves how necessary decentralised research and participation by the farmers really is. However, it also demonstrates the problems of this approach: the authorities, the agricultural research institute and the extension officers are constantly in opposition because they are anxious that they might lose control over the developments. The search for solutions in pest control which link up with the needs of farmers' co-operatives must nevertheless be carried out in collaboration with the farmers. For this, encouragement should be given by the authorities. As long as these, the extension officers and researchers are not prepared to devote their energies to assist the interests of the small farmers, they leave little opportunity for the proper research and applications needed for this group. Unfortunately, the Peruvian government is exclusively interested in higher production for a number of economically important crops, and therefore the introduction of IPM into Peru remains an apparently impossible dream in the foreseeable future. An IPM program has little chance of success as long as the mass production of a few crops, with enormous investment in artificial fertilizers and pesticides, is given preference.

The small-scale action by field-workers, such as the development of a more environmentally sound control of pests in maize, is the first step. However, it is important that the Peruvian government is aware of the dangers of the large-scale use of pesticides and for this purpose pressure must be exerted from both above and below to bring this to their attention: both from the international organizations as from the farmers themselves.

Appendix I
Safe use of pesticides, plus
Pesticide list

I.1 Help in case of poisoning

If somebody suffers from poisoning the following rules must be followed:

- **Immediately call for medical assistance or take the victim to a doctor.** Notify the doctor of the active ingredient contained in the pesticide. While waiting for medical assistance to arrive take the following measures:
- **Remove the victim from the source of poisoning.**
- **Check respiration and make sure that the airway is clear.**
- **If spontaneous breathing is inadequate give artificial respiration.**
- **Remove contaminated clothing in contact with the skin** and wash the body carefully with soap and water. Do not do this if the skin is actually damaged.
- **Keep the victim calm.** Lay him/her out of the sun either in the shadow or in a covered spot. If poisoning is by a pesticide which causes respiratory problems (e.g. DNOC, dinoseb, binapacryl or pentachlorophenol) lay the victim in a cool place.
- **If an eye is contaminated: rinse immediately** with clean, running water for at least ten minutes. Make sure the other eye is not touched. Help the patient to keep the eye open and allow the water to stream gently over the eyeball.
- The oral intake of poison must be treated rapidly, particularly if the pesticide swallowed was in an aquatic solution, the victim should be made to rinse the mouth thoroughly a few times and to drink at least two glasses of water. Vomiting should be subsequently induced e.g. the victim should drink a salt water solution (two dessert-spoons of salt in a glass of warm water), or the back of the throat should be irritated. After vomit-

ing, the victim should be given a strong dose of charcoal (either a number of finely powdered tablets or a dessert-spoon of powders in a glass of water).

If the pesticide swallowed was diluted with an organic solution (e.g. kerosine) the victim should be given a large quantity of beaten egg or a starch solution before vomiting is induced. In the event of caustic chemicals, acids or lixiviates, vomiting should never be induced but a charcoal powder should be given.

- **If convulsions should occur, prevent the patient from injuring him/herself.** One important measure is to place a well-folded handkerchief between the teeth to prevent biting of the tongue. Take care of your own fingers in such an event.

If the victim has lost consciousness:
In the absence of medical assistance for unconscious patients little else can be done than to ensure that the victim is able to breathe freely. Lay the patient on his side in a stable position and make sure that the air-passages are not blocked by tilting the head backwards with the mouth open; remove any vomit, food or false-teeth and pull the tongue to the front. Always remove any restrictive clothing.

Never induce vomiting or give any drink in case a victim is unconscious, these could cause asphyxiation. If necessary, clean the eyes and skin of the patient (see above) and give artificial respiration – making sure to avoid any contact with the poison.

In some cases an antidote can provide relief, however, this should be administered preferably by a doctor.

I.2 Storing pesticides

- **Store the pesticides in special areas which can be properly closed.** The area must be dry and preferably cool and furnished with proper ventilation and lighting.
- **Storage areas for pesticides must never be used for keeping any other commodities e.g. food, fodder, clothing, tools etc.**
- **Do not enter the storage area alone.**
- **Never leave pesticides unguarded outside**

the storage area.
- **Empty packaging must not be left lying about and must be disposed of properly.** Where possible, empty packaging should be burnt, metal containers can be buried. Re-using metal containers for storing drinking water is a common but irresponsible practice.
- **Always store pesticides in their original packaging!** Even small amounts left over; never

transfer pesticides to other bottles such as coca-cola or beer bottles. The labels from the original packaging should be kept.

I.3 Safe use of pesticides

In warm weather the chance of poisoning by pesticides is greater than in cool weather. The flow of blood in the skin increases with higher temperatures and therefore pesticides are more easily absorbed. Protective clothing then is also less comfortable to wear. In Third World countries therefore pesticides should be treated with extra special care.

I.3.1 General rules for the safe use of pesticides

Always observe the following rules of conduct:

- **Always read the label thoroughly before using the pesticide.** Never use pesticides which are not properly labelled. The label indicates what safety measures should be taken when applying the pesticide, for example: "Special Hazards", "Warning", "Safety Recommendations" or 'Precautionary Measures' and also describes the first aid treatment in the event of poisoning. See fig. 15.1.
- **Take extra precautionary measures when mixing pesticides**, then you are working with high concentrates. Use proper pouring equipment and funnels and wear protective clothing, at very least impermeable gloves. Prevent powders from being blown around when mixing.
- **Before use, check spraying equipment for any leakages.** Handle spraying equipment with care. **Never** use leaking equipment.
- **When spraying make sure that water, soap and towels are available** so that the skin can be washed immediately in the event of contamination. After spraying wash the whole body and change clothing.
- **When preparing the spraying solution: do not eat, do not drink or smoke** and do not do so until spraying is completed and hands have been washed thoroughly. Smoking and alcoholic drinks before, during or after spraying are an absolute evil since they cause physiological changes in the body which can increase the toxicity of a pesticide.
- **Be aware of the possibility of poisoning**

- **Keep pesticides away from children and animals.**

cattle, fish game and birds. Do not spray pesticides which are toxic to bees on flowering crops.
- **Avoid over-long working days when spraying.**
- **Never rinse spraying equipment in surface water.**
- **Keep away from recently sprayed crops.**

I.3.2 Protective clothing

The best way to avoid poisoning is to avoid physical contact with chemicals. It is often difficult to convince people about the necessity to wear protective clothing since these are uncomfortable to wear, particularly in warm weather, while the injurious effects of pesticides are often not immediately felt or visible. Make sure, however, that at least if the labelling states that protective clothing should be worn, these directions are followed.

- **Jacket and trousers**; if the user can become wet during spraying, an overall with an impervious apron should be worn. The clothing should offer the body complete and good coverage. Most important is that clothing worn during spraying is absolutely impervious to the spraying solution.
- **Gloves**; because hands frequently come into contact with the pesticides, on buckets, bottle-tops, etc. rubber or synthetic gloves must be worn. Talcum powder or a lining in the gloves make them more comfortable when transpiring. The inside of the gloves must be cleaned regularly to ensure than any pesticide which has got inside does not come into continual contact with the skin.
- **Rubber boots**; trousers should be worn over the boots to avoid the spraying solution dripping in. Wash the boots after spraying both inside and out.
- **Protective glasses or face-shield**; to prevent the eyes being splashed glasses or, even better, a face-shield must be worn.
- **Hat**; a hat or cap will protect hair and face from a dusty pesticide.

The above recommendations are made on the assumption that all protective clothing is well-labelled, which in many Third World countries is often not the case. However, always try to realise the highest degree of safety with whatever means are available.

I.3.3 Recommendations for safe use of pesticides based on their toxicity.

The WHO classified pesticides on which recommendations for safe use are based. Pesticides are classified according to their mammal toxicity (median lethal dose (=LD-50) for rats) and physical state (fluids are more hazardous than solids). In fact, only active ingredients are classified although the hazard level is also influenced by formulation, way of application etc.

Classes and recommendations for use are:

Class Ia. Extremely hazardous pesticides. Should be applied only by individually licenced operators.
When handling the concentrate or applying in the field, wear: boots, an overall, a hat, gloves, a face shield or mask, eventually an apron.

Class Ib. Highly hazardous pesticides. Should be applied only by trained and supervised operators.
When handling the concentrate, wear: boots, an overall, a hat, gloves, a face shield or -mask.
When applying in the field, wear: canvas shoes, an overall, a hat and a face mask.

Class II. Moderately hazardous pesticides. Should be applied only by trained and supervised operators and in strict observance of the prescribed precautionary measures.
When handling the concentrate, wear: boots, an overall, a hat, gloves, a face shield or -mask.
When applying in the field, wear: canvas shoes, an overall, a hat and a face mask.

Class III. Slightly hazardous pesticides. Should be applied only by trained operators and in observance of the routine precautionary measures.
When handling the concentrate, wear: boots, an overall, a hat, gloves, a face shield or -mask.
When applying in the field, wear: canvas shoes, an overall and a hat.

Class III. Pesticides which are unlikely to present a hazard in normal use. However, it is possible that in formulations, solvents or vehicles the hazards may be greater than with the actual pesticide. These pesticides should therefore be used under the same precautions as pesticides from class III.

I.4 Description of the pesticides

The following lists a number of pesticides frequently used in the Third World and gives a description of some of their characteristics.

ACTIVE INGREDIENT, indicates the common name of the active ingredient.

Important trade names: indicates some of the trade names of commercial formulations. Of course, it is not possible to give a complete list of all the trade names.

Properties: names some chemical and physical properties of the chemical. 1. Chemical group. 2. Appearance at room-temperature. 3. Melting point. 4. Solubility in water at 20-30 °C. 5. Stability. 6. (if known) keeping qualities.

Mode of action: 1. Required form of contact with the target organisms for effective action. 2. Systemic/non-systemic pesticide. 3. Action mechanism in the target organisms.

Use: 1. Target organism. 2. Crops in which the pesticide may be used. 3. (if available) prescribed dosage of the chemical for normal use.

Toxicity: WHO-classification for toxicity of the chemical.
Short term: 1. Lethal Dose at which 50% of the testing animals died (LD-50) in mg active ingredient of the pesticide per kg body weight of the testing animals, as found in literature. 2. Information on human toxicity of the chemical.
Long term: 1. No-Effect-Level (NEL) in mg active ingredient per kg food for rats, in feeding trials of 6 months to 2 years. 2. Fate of the chemical in the body of mammals. 3. Information on toxicity of the chemical to mammals.

Safe use: 1. Safe use of the WHO-classified product. 2. Special information on safe use of the chemical. 3. Period a treated field must not be entered.
Environmental effects: 1. Global toxicity of the

chemical for fish, birds and bees. 2. Fate of the chemical in the environment (accumulation, half-life).

Registration: 1. Admitted in the EC or USA. 2. Admitted in Third World countries, if known.

ALDRIN, DIELDRIN, ENDRIN   insecticide

Trade names: Aldrin: Octalene. Dieldrin: Dieldrex, Dieldrite, Octalox. Endrin: Endrine, Nendrin.
Properties: The three insecticides are organo-chlorines. Aldrin is a tan to dark brown solid m.p. 49-60 °C, solubility mg/l water. It is stable in a neutral, basic and weakly acidic media, but it reacts with strong acids and phenols in the presence of oxidizing agents to give dieldrin.
Dieldrin consists of buff to light tan flakes with a mild odour, m.p. 175 °C, solubility 0.186 mg/l water. Dieldrin is stable to alkali, mild acids and to light, but sensitive to strong acids and some carriers. Endrin is a light tan crystalline solid, m.p. 50-60 °C, solubility in water is negligible. It is stable to alkalis but unstable to strong acids.
Mode of action: As other organochlorines (see DDT). It acts by contact and ingestion.
Uses: Broad spectrum insecticides used against many pests in a wide variety of crops. Some target organisms: beetles, termites, locusts, caterpillars and butterflies. Some crops: cotton, tobacco, sugar-cane, peanuts, rice, jute, maize, soya. Also used to treat seeds to prevent insect damage. Also used against various vectors such as the malaria mosquito and the tse-tse fly. Aldrin is used in particular against soil-pests. Most commonly applied doses: aldrin: 0,5-5 a.i kg/ha; endrin 0,2-0,5 kg a.i./ha.
Toxicology: WHO-classification: Ib.
Short term: Aldrin, endrin and dieldrin are all acutely toxic pesticides. Oral LD-50 of aldrin is 38-60 mg/kg, dermal LD-50 98 mg/kg. Oral LD-50 of dieldrin is 46 mg/kg, of endrin 7-18 mg/kg, percutaneous LD-50 of endrin is 15 mg/kg (all for rats). Orally taken, inhalation and absorbtion through the skin can result in poisoning. Sensitivity is increased by malnutrition. Symptoms of poisoning are the same as those of organochlorines (see DDT). Repeated overexposure to aldrin or dieldrin may result in convulsions, sometimes days after exposure has ceased and without any signs of previous symptoms, and bone fractures and tongue-biting are not uncommon.
Long term: As for organochlorines (see DDT). Various experiments with animals have proved the carcinogenity of dieldrin. About possible long-term effects of aldrin and endrin little is certain. It is suspected that these pesticides can lower the fertility of females. In the body aldrin is converted to the more dangerous dieldrin.
Safe use: For precautionary measures see I.3.3 class Ib. In the event of poisoning treat as for organochlorines (see DDT). Keep the pesticide away from water. Endrin-poisoning should be treated with barbiturates.
Best advice for these pesticides is NOT to use them, especially not on food crops and never by pregnant women.
Environmental effects: In moderate regions the breakdown of the drins, as for most organochlorines is very slow; as a rule a over a 3-8 year period approximately 95% is broken down or disappears. In tropical conditions break down is generally slower, caused by rapid volatilization in the air. In exceptional circumstances (i.e. very dry or wet conditions) breakdown in the tropics can take up to 20 years. The drins are a hazard for bees and fish (therefore of high risk in irrigated rice fields!), for birds, especially those that eat off insects and fish, and for many mammals. The pesticides accumulate to high levels in the food chains. The use of drins can also seriously damage the build-up of natural enemies of insects.
Registration: Prohibited or restricted in many countries (EEC-countries) because of the toxicity aspect and because of persistence in the environment, accumulation in food chains, resistance problems and the frequently high residues in food and milk etc.

ALDICARB — insecticide, nematicide

Trade names: Temik, Temizid, UC 21149.

Properties: Chemical class: carmaboyl-oximes. Aldicarb forms colourless crystals. The melting-point is at 98-100 °C , it breaks down above 100 °C, solubility 6 g/l water. Aldicarb is stable except to strong alkalis and is sensitive to heat. With oxidizing agents is converted rapidly to sulphoxide and converted slowly to sulphone.

Mode of action: It is a contact poison with systemic action. Intake through the roots is rapid and the plants are protected for up to three months.

Uses: Aldicarb is a soil-applied systemic pesticide used against certain mites, nematodes and insects (especially aphids, whiteflies, leaf miners). Seed furrow, band or overall treatments at 0.56-11.25 kg a.i./ha are used. Soil moisture is required to release the active chemical from the granules, so irrigation or rainfall should follow application.

Toxicology: WHO-classification: Ia.

Short term: The acute oral LD-50 for male rats is 0.93 mg/kg. The acute percutaneous LD-50 for male rabbits is 5.0 mg/kg. Rats were killed within 5 minutes by a dust concentration of 200 mg/m^3 air. Therefore aldicarb is one of the strongest poisons for mammals.

Long term: NEL for rats 0.3 mg/kg diet.

Safe use: For safe use see I.3.3 class Ia. Strictly observe general rules for the use of pesticides. Carefully avoid contact with eyes and skin and inhalation of the vapour. Do not mix granules with water. Do not use applicators that will grind the granules. Wash contaminated clothing in strong washing soda solution and rinse thoroughly.
In the event of oral intake, induce vomiting immediately. Keep the patient calm until medical help arrives.

Registration: Aldicarb is usually admitted in ornamentals and some food crops (bulb onions, tomatoes, brassicas, peas, potatoes, maize, soya beans, citrus, bananas, coffee, sorghum and sugar cane). Treated fruit may not be consumed in the year of treatment.

ATRAZINE — herbicide

Trade names: Gesaprim, Primatol, AAtrex, Atranex, Atred, Vectal, Actazin, Aktikon, Argesin, Atazinax, Atrozol, Zeazin.

Properties: Chemical group: Triazines. Atrazine is a clourless powder. The m.p. is at 175-177 °C, solubility in water at 20 °C is 30 mg/l. It is stable in neutral, weakly acidic and weakly alkaline media. It is hydrolyzed to the herbicidally-inactive hydroxy derivative in acids and bases at higher temperatures and is very stable over several years in storage.

Mode of action: It interferes with photosynthesis and other enzymic processes in the plant. In tolerant plants it is metabolized readily to hydroxyatrazine and amino acid conjugates, with further decomposition.

Uses: Atrazine is a pre- and post-emergence herbicide used in maize, asparagus and sweet corn. It is a soil and leaf herbicide and absorbed principally through the roots, but also through the foliage. It is used to control of grass under young forest trees, fruit trees and vines. Used at high application rates to control weeds non-selectively in non-cropped areas. Also used in pineapple, sorghum and sugarcane.

Toxicology: WHO-classification: III.

Short term: Acute oral LD-50 for rats is 3080, for rabbits 600 mg/kg. The acute dermal LD-50 for rabbits is 7500 mg/kg. The inhalation LC-50 for rats is more than 710 mg/m^3 air (1 hour). Atrazine is slightly irritant to skin and eye.

Long term: Not known.

Safe use: For safe use see I.3.3 class III. Wash after use and avoid skin and eye contact and inhalation of dust and spray mists. Guard against drift. Do not contaminate food, fodder and water supplies. Atrazine is metabolized and excreted by mammals quite rapidly.

Environmental effects: Atrazine has a half-life in soil of about 3 months. Residual activity prevents the growing of some (sensitive) crops.

Registration: 1990 most EEC countries will ban atrazine.

BENOMYL fungicide

Trade names: Benlate, Tersan 1991.
Properties: Chemical group: benzimidazoles. Pure benomyl is a colourless crystalline solid; on heating it breaks down without melting, solubility 2 mg/l water. Benomyl breaks down if stored in contact with water and under moist conditions in soil.
Mode of action: It is a systemic fungicide.
Uses: It is a protective and eradicant fungicide with systemic activity and is effective against a wide range of fungi affecting field crops, fruits, nuts and ornamentals. It is also effective against mites, primarily as an ovicide. It is used as pre- and post-harvest sprays or dips to control storage rot of fruits and vegetables. Typical rates are: on field and vegetable crops, 140-550 g/ha; on tree crops 550-1100 g/ha.
Toxicology: WHO-classification: III.
Short term: Acute oral LD-50 for rats 10,000 mg a.i./kg. Acute dermal LD-50 for rabbits 10,000 mg/kg.
Long term: None.
Safe use: For safe use see I.3.3 class III. Excercise the general precautions.
Environmental effects: Half-life in soil 6-12 months.
Registration: Benomyl is admitted widely.

CAMPHECHLOR insecticide, rodenticide

Trade names: Alltox, Anatox, Chemphene, Clorchem T-590, Toxakil, Toxaphene.
Properties: Camphechlor is a mixture of organochlorines. It is a yellow wax of mild terpene-like odour, softening in the range 70-95 °C, solubility 3 mg/l water. It is dehydrochlorinated by heat, by strong sunlight and by certain catalysts such as iron.
Mode of action: It is a non-systemic contact and stomach insecticide with some acaricidal action. It is also a rodenticidal stomach poison.
Uses: It is used in the control of many insects, grasshoppers, armyworms and cutworms, on corn, cotton, fruit, small grains, vegetables and in soyabeans. It is also used for the control of animal ectoparasites, horn flies, lice, ticks etc.
Toxicology: WHO-classification: II.
Short term: The acute oral LD-50 for rats is 80-90 mg/kg. The acute dermal LD-50 for rats is 780-1075 mg/kg. The acute toxicity is high and some cases of poisoning have been known to be fatal. The concentrate is readily absorbed by the skin and the dust can be inhaled. Symptoms are the same as poisoning by other organochlorines.
Long term: NEL for rats is 25 mg/kg diet. Accumulation in the body fat is proportional to the dose, but elimination proved to be rapid when intake was stopped. Probably the mixture is carcinogenic and it affects genetic material.
Safe use: For safe use see I.3.3 class II. Exercise the general precautions. For properties of organochlorines see DDT.
Camphechlor in fact is chemical waste. It is in fact not possible to work safely with a pesticide such as camphechlor, and protection against long-term effects is more or less impossible to effect.
Environmental effects: Camphechlor is hazardous for the environment. Acute toxicity to water-organisms and birds is high. The persistent character of the pesticide causes it to accumulate in the food chains. Camphechlor is mobile and is able to disperse over wide areas. See further organochlorines (DDT).
Registration: Camphechlor is prohibited or severely restricted in many European countries.

CAPTAN fungicide

Trade names: Altan, Captane, Flit 406, Merpan, Orthocide 406, Orthocide, Trimegol-50, Vancide 89, Vondocaptan.
Properties: Captan forms colourless crystals, m.p. 178 °C, solubility 3.3 mg/l water. It is unstable under alkaline conditions.
Mode of action: It is a curative leaf fungicide.
Uses: It is a fungicide to control diseases of many fruit, ornamental and vegetable crops. It should not be mixed with oil sprays. It is used as a spray, root dip or seed treatment to protect young plants against rot and damping-off.
Toxicology: WHO-classification: III.
Short term: The acute LD-50 for rats is 9000 mg/kg. It may cause skin irritation.
Long term: NEL for rats was 1000 mg/kg diet. No teratogenic or mutagenic effects have been observed.
Safe use: For safe use see I.3.3 class III. Use a respirator when working with the concentrate. Avoid pro-

longed contact with the skin.
Environmental effects: Toxic to fish.

CARBARYL   insecticide

As carbaryl is one of the main carbamates in use,
some generalities concerning this group of insecticidal
compounds are given.
Carbamates.
As organophosphorous compounds, carbamates are
anticholinesterases. Carbamates are all direct inhibitors
of cholinesterase (they do not have to be converted
into an active compound first), so their
action is quite fast. They are quite poisonous to insects
and mammals, causing reaction in the brain,
glands and muscles. Symptoms of poisoning are:
headaches, dizziness, nausea, vomiting, blurred vision,
increased sweating, dribbling at the mouth,
tremor, muscular twitchings, etc. Symptoms usually
occur rapidly, causing an inability to continue working.
As a rule carbamates are readily absorbed, metabolized
and excreted by mammals. Their action and
symptoms are of short duration, shorter than after
poisoning with organophosphorous compounds.
Symptoms of poisoning can be treated by injecting
1-2 mg atropine (-sulphate). Atropine pills do not
have a detoxifying effect but they do relieve the
symptoms somewhat. Sufferers of liver and kidney-diseases
must not work with carbamates.
Trade names of carbaryl: Denapon, Dicarbam,
Murvin, Patrin, Ravyon, Sevin.
Properties: Carbaryl is a carbamate. It is a colourless
crystalling solid, m.p. 142 °C, solubility
120 mg/l water. It is stable to light and has also a
wide stability in neutral and weakly acidic conditions.
It is hydrolyzed in alkaline media.
Mode of action: Carbaryl predominately affects
the stomach, and also has some contact-reactions.
It acts as all carbamates.
Uses: It is a non-selective pesticide used to control
many insects in particular in cotton, rice, maize,
fruit, potatoes and vegetables. Also used on cattle to
control ticks, lice and fleas. In de US it is the most
commonly used insecticide. The recommended dose
is 250-2,000 g/ha.
Toxicology: WHO-classification: II.
Short term: Carbaryl is a fairly toxic pesticide, especially
when taken orally (oral LD-50 in rats 400-850

mg/kg). It can also be absorbed through the skin
(dermal LD-50 in rats is more than 4000 mg/kg)
and inhaled. Symptoms of poisoning can occur
with an intake of a quarter gram (see the symptoms
for carbamates). A protein deficient diet can significantly
increase toxicity. Because the pesticide is excreted
fairly rapidly after poisoning (in the urine)
recovery is fast.
Long term: Carbaryl does not accumulate in the
body. Research into the mutagenity and carcinogenity
of carbaryl revealed no negative effects in its
use. However, dogs and chickens showed indications
of birth-defects after lengthy exposure. Animal
experiments also showed that lengthy exposure
caused a reduction in fertility in both males and females.
Safe use: For safe use see I.3.3 class II. In the event
of accidental intake: cleanse the stomach with a
soda-solution. See further carbamates.
Environmental effects: Carbaryl in the environment
has a short half-life. As a result of microbial
breakdown after approximately 4 months it has as
good as disappeared from the soil. In water carbaryl
can temporarily halt the growth of algae and in consequence
disturb the natural balance. Acutely toxic
to bees. Toxicity to birds and fish is slight.
Registration: Carbaryl is prohibited nowhere.

CARBOFURAN   insecticide

Trade names: Furadan, Curaterr, Yaltox.
Properties: Carbofuran is a crystalline solid, m.p.
150-152 °C, solubility at 25 °C 700 mg/l water, it is
essentially insoluble in conventional formulation
solvents used in agriculture. It is unstable in alkaline
media.
Mode of action: Carbofuran is a carbamate. It has
stomach- and contact action.
Uses: Carbofuran is a systemic acaricide, insecticide
and nematicide, applied to foliage at 0.25-1.0 kg
a.i./ha for the control of insects and mites it is applied
to the seed furrow at 0.5-4.0 kg/ha for the control
of soil-dwelling and foliar-feeding insects, or
broadcast at 6-10 kg/ha for the control of nematodes.
It is used in maize, brassicas and rice to control
flea beetles, root flies, stem weevils, the potato
cyst nematode etc.
Toxicology: WHO-classification: Ib.
Short term: Acute oral LD-50 for rats 8-14 mg/kg.

Can be absorbed through the skin.
Long term: In 2-y feeding trials no effect was found on rats receiving 25 mg/kg diet. It is metabolised in the liver and excreted in the urine of animals, 50% being lost in 6-12 h.
Safe use: For safe use see I.3.3 class Ib. Scrupulously avoid inhalation and skin contact. The precautions are the same as for carbamates (see under carbaryl). If swallowed, drink one or two glasses of water and induce vomiting. If coming into contact with the eyes, administer one drop of homatropine.
Environmental effects: Carbofuran is very toxic to fish, and moderately toxic to birds. In soils 50% is lost in 30-60 d.
Registration: Prohibited or severely restricted in EEC-countries.

CHLORDANE   insecticide

Trade names: Belt, Octachlor.
Properties: Chlordane is an organochlorine. It is a viscous amber liquid and practically insoluble in water. It is sensitive to alkalis and under UV irradiation a change in the structure occurs.
Mode of action: It is a non-systemic stomach, contact and respiratory insecticide with some fungicidal action. The mode of action is like other organochlorines (see DDT).
Uses: This non-selective insecticide is used on land against ants, coleopterous pests, cutworms, grasshoppers, termites, cockroaches and many other insect pests. It also controls pests on humans and domestic animals and is used as a wood preservative. It may be applied to soil, directly to foliage or as seed treatment.
Toxicology: WHO-classification: II.
Short term: Oral LD-50 for rats is 460-600 mg/kg, dermal LD-50 200-2,000 mg/kg. Can cause poisoning if taken orally, through the skin or by inhalation. Intake of approximately 6 grammes can be fatal. Many fatal cases of poisoning are known many of which due to chlordane spilt on the skin. People with liver-afflictions (alcoholics) are known to be especially sensitive. The symptoms of poisoning are very similar to those of DDT.
Long term: The NEL in rats is 60 mg/kg diet. The accumulation-levels of organochlorines is strongly suspected to cause deformity of liver tissue. Liver-afflictions are given the same priority as accumula-

tions. Experiments with animals have indicated that chlordane is carcinogenic, in particular with regards to the liver. Chlordane is also suspected to have teratogenic characteristics.
Safe use: For safe use see I.3.3 class II. A detoxifying agent is not known. It is advisable to avoid totally working with this pesticide. Unless sufficiently protected, people should keep away from treated areas for at least 24 hours.
Environmental effects: Chlordane is one of the most cumulating pesticides (in the form of the carcinogenic heptachlor-epoxide) out of the organochlorine group (see under DDT). It accumulates in the food chains. The half-life varies from 1 to 14 years according to the circumstances. A good illustration of the high level of persistency is the fact that wood which was treated with chlordane in the first year of application (1945) still shows sufficient resistence to termites. In water chlordane disappears faster, this is especially due to evaporation. Chlordane is toxic to water organisms, birds, mammals and bees.
Registration: Chlordane is banned in many countries. In a number of countries it is admitted only for use against soil-pests.

CHLOROBENZILATE   acaricide

Trade names: Akar, Folbex, Acaraben, Benzilan, Clobex, Kopmite, Pomite.
Properties: Chlorobenzilate is a colourless solid, m.p. 36-37.5 °C; solubility at 20 °C: 10 mg/l water.
Mode of action: It is a non-systemic acaricide with contact action.
Uses: Chlorobenzilate is an acaricide with little insecticide action. It is recommended for use against phytophageous mites on citrus, cotton, grapes, soyabeans, tea and vegetables at 1.0-1.5 kg/ha.
Toxicology: WHO-classification: III.
Short term: Acute oral LD-50 for rats 3000 mg/kg. Non-irritant or toxic to skin.
Long term: NEL for rats in 2-year feeding trials 40 mg/kg diet.
Safe use: For safe use see I.3.3 class III. Observe the normal precautions; avoid contact with the eyes and skin and do not inhale the spray mists.
Environmental effects: Moderately toxic to fish. Practically non-toxic to birds and honeybees.

Registration: Forbidden in some EC-countries because of insufficient environmental information.

CHLORPYRIFOS

Trade names: Detmol, Dursban, Dwco 197, Edil CP, Lorsban, Loxiran, Pyrinex, Zidil.

Properties: It is an organophosphorous compound, forming colourless crystals with a mild mercaptan odour, m.p. 43 °C, solubility at 25 °C: 2 mg/l water. It is compatible with non-alkalin pesticides but corrosive to copper and brass.

Mode of action: It has a non-systemic contact, stomach and respiratory action. It is absorbed through the leaves and roots and it has a slight translocation.

Uses: Chlorpyriphos has a broad range of insecticidal activity and is effective by contact, ingestion and vapour action, but is non-systemic. Used for the control of flies, household pests, mosquitoes (larvae and adults) and of various crop pests in soil and on foliage; also used for control of ectoparasites on cattle and sheep. Its volatility is great enough to form insecticidal deposits on nearby untreated surfaces. It is non-phytotoxic at insecticidal concentrations.

Toxicology: WHO-classification: II.

Short term: Acute oral LD-50 for rats about 150 mg/kg

Long term: It is rapidly detoxified in rats, dogs and other animal species.

Safe use: For safe use see I.3.3 class II. Avoid contact with eyes and skin and the inhalation of the vapours, dust or sprays. It must not be used in the immediate vicinity of water. In other respects, see parathion.

Environmental effects: It is degraded in soil, in 60-120 days. It is toxic to fish, birds and to bees.

Registration: Forbidden or severely restricted in many European countries.

COPPER COMPOUNDS   fungicides

Trade names: Copper sulphate ($CuSO_4$) is also known as vitriole-blue, among others. Other copper compounds used: copper-carbonate copper-oxychloride, copper-naftenate, copper-hydroxide, copper-oxy-chinolate.

Properties: These are all compounds with the copper-ion as active component. Blue or green crystals.

Mode of action: The copper component of these compounds is responsible for their fungicidal effect. It stops germination of fungal-spores as well as the growth of hyphe. Copper may injure the crop.

Uses: Applied as fungicide, formerly in common use.

Toxicology: Copper sulphate is classified in WHO-class II.

Short term: Of all the copper compounds, copper-carbonate is the most and copper-oxychloride the least toxic. We discuss here the effects of especially copper-sulphate. This is a fairly toxic fungicide, its intake is normally oral or through inhalation. It can cause great irritation of the eyes, skin and mucous membrane, and can give pain, redness and can disturb vision if coming in contact with the eyes. It will induce vomiting if swallowed. Poisoning occurs with an intake of approximately 1 gram, 8-15 grammes can be fatal.
Symptoms of poisoning: metalic taste, vomiting of a blue/green lump, heavy watery and bloody diarrhoea, dehydration and possibly shock. Acute poisoning will cause victims to become seriously ill and to die within a few hours. If victims survive they are likely to suffer heavy liver, kidney and blood afflictions.

Long term: If absorbed, the copper component of the pesticide accumulates in the body and ends up particularly in the liver. No literature makes any mention of negative effects on fertiliity or of carinogenity.

Safe use: For safe use see I.3.3 class II. Always work with gloves and face-shields. In the event of poisoning: call for medical assistance, victim should rinse mouth thoroughly with water and then drink plenty of water or milk. If available treat victim with 1% potassium ferrocyanide and open bowels with sodium sulphate.

Environmental effects: Used excessively and over long periods the heavy metal, copper, accumulates in the soil. A number of localities in France are now suffering from soil-fertility problems; the effect of having used copper containing fungicides in the vineyards for almost a hundred years.

Registration: Copper compounds are admitted as fungicides almost everywhere.

2,4-D   herbicide

Trade names: BSI, ISO, WSSA.

Properties: Chemical group: phenoxy acetic acids. Pure 2,4-D forms colourless crystals. The ion forms various salts, with a melting point from 85 to 159 °C, solubility 60 mg/l water, some salts up to 4,4 kg/l water.

Mode of action: Disturbs the natural plant hormone balance in plants causing uncontrolled growth.

Uses: 2,4-D is a herbicide used against wide leaved weeds; grass-like weeds usually are not sensitive to it. For this reason it is commonly used in cereals. Other main crops are: maize, sugar-cane, grass-land, forestry. Often available in mixtures.

Toxicology: WHO-classification: II.

Short term: Lowest LD-50 for rats 375 mg/kg. 2,4-D and other hormonal herbicides herbicides (MCPA, MCPP and 2,4,5-T) are moderately toxic to warm-blooded animals. It is available as a free acid and as a salt. It is most hazardous in acid form. The pesticide is particularly toxic if taken orally or inhaled, and to a lesser degree if absorbed through the skin. Symptoms of poisoning: vomiting, headaches, double-vision, loss of bladder control, slack muscles and in serious cases unconsciousness (coma). Convulsions may also occur. Used in combinations with certain insecticides such as malathion and parathion, toxicity increases.

Long term: NEL for rats 625 mg/kg food. 2,4-D does not accumulate in fat or other tissue. Breaks down and is excreted quite rapidly, mainly through the liver and kidneys. Regular exposure to large doses 2,4-D can negatively effect these organs. Reports about genetic and carcinogenic characteristics and such like are contradictory.

Safe use: For safe use see I.3.3 class II. Avoid continuous long-distance exposure to small amounts. Prevent drifting. Should symptoms of poisoning appear, cease work, remove clothing wash affected areas thoroughly. In the event of accidental intake: induce vomiting, call for medical assistance, and rinse stomach with 5% sal volatile. Artificial respiration may be necessary (in case victim is in a state of coma). Water contaminated with 2,4-D must not be used as for domestic or any other purposes.

Environmental effects: 2,4-D break-down in soil is fairly rapid, its half-life is 1-2 weeks. Moderately toxic to fish and birds. Does not accumulate in the food chains. Treated fields should not be set to grazing for two weeks after spraying. Is not toxic to bees. 2,4-D inhibits nitrification in wet rice fields.

Registration: Because of its effects on the environment India has totally prohibited the pesticide, in Colombia it is only prohibited for use in coffee.

DDT   insecticide

DDT is one of the most commonly known and notorious pesticides. In 1949 the discoverer of DDT was awarded the Nobel prize; two decades further on measures were taken everywhere to restrict its use. At least two million tons of DDT have been applied since the forties, and in spite of the fact that it is recognised as having many undesirable characteristics, the production and use of DDT in various countries simply continues. DDT is an organochlorine, one of the main pesticidal groups. The following starts with some general details about organochlorines and then goes on to deal with DDT specifically.

Organochlorines are insecticides usually quite persistent in the environment. Insects often develop resistance to this class of compounds. The major toxic action is on the nervous system. Organochlorines tend to vary considerably in their toxicity to mammals. They are soluable in fat and can be therefore stored in the body fat with no apparent effect. Their persistence, however, causes considerable health and environment problems. The cumulative effect of chlordane DDT lindane, camphechlor methoxychlor.

Toxicity. Acute poisoning from organochlorines is rare, unless massive exposure occurs. The action of this group of insecticides is confined to the central nervous system, and the symptoms are those of CNS stimulation. Within a few hours after ingestion symptoms are: headache, apprehension and excitement along with dizziness, followed by disorientation, vomiting, muscular weakness, tremors and finally convulsions. Respiration is initially accelerated but later fails and the victim may die. When organochlorines are absorbed through the skin, apprehension, mental confusion and tremors may be the only symptoms. Small doses may cause anorexia.

Treatment of poisoning. There is no specific

antidote, but supportive and symptomatic treatment may be life-saving. Clear the air-passages. Ensure the patient cannot self-inflict injury when convulsions occur. If the pesticide was swallowed, induce vomiting and rinse the throat with a *non-oily* purgative. Do not use milk: organochlorines dissolve in fat and are absorbed in that way. If the skin and eyes are contaminated they should be washed thoroughly. Call a doctor.

If the victim survives and there has been no brain-damage, recovery is usually complete.

Major trade names for DDT: Dedetane, Didimac, Gesapon, Gesarol, Neocid.

Properties: DDT is an organochlorine. It forms colourless crystals with a m.p. at 109 °C. It is practically insoluble in water and above 50 °C not very stable. Iron, aluminium and UV-light promote decomposition.

Mode of action: DDT is a potent non-systemic stomach and contact insecticide. It is a nerve-poison which acts to stimulate the central nervous system. There is insufficient information available about its precise effect at cellular level.

Uses: DDT is a broad-spectrum pesticide used particularly on cotton, cabbage and maize. DDT together with malthion are the most commonly used pesticides in the control of the malaria mosquito. DDT is still used in abundance in campaigns to control the tse-tse fly.

Toxicology: WHO-classification: II.

Short term: The oral LD-50 for rats is 115 mg/kg, the dermal LD-50 with female rats is 2510 mg/kg. DDT can enter the body through the mouth and through the air-passages. If dissolved in oil DDT can easily be absorbed through the skin. Symptoms of poisoning appear at doses of above 6 mg/kg. The fatal dose for humans is 60-500 mg/kg. Bad nutrition, particularly protein-deficiency, increases sensitivity. Symptoms of poisoning are the same as those for organochlorines in general.

Long term: There have been reports that DDT is carcinogenic but these are not clear. There are no definite indications of mutagenity; the cause of birth defects, and it does not affect fertility in humans. DDT is stored largely in the body fat, which is able to contain up to 100 g/kg. DDT can enter the body via the food chains, and especially with meat and milk products. Moreover, pregnant women can pass it on to their babies through the placenta or via their milk if breast-feeding.

Safe use: For safe use see I.3.3 class II. Protective clothing is needed to prevent skin-contact and inhalation. A specific antidote in not known. If the poisoning is not fatal, recovery within a few days can be expected. People suffering from liver or kidney-afflictions are strongly advised not to work with DDT. Treated fields should not be entered for at least 24 hours after spraying without sufficient protection.

Environmental effects: Chemically and biologically DDT is a very stable compound even when exposed to heat, acid, UV rays (sunlight) and micro-organisms. In temprrate climates the half-life is 3-10 years. An unknown amount 'disappears' due to evaporation. Breakdown by micro-organisms occurs especially in anaerobic conditions but is faster in tropical soils: an estimated 90% disappeares within a year. These figures only refer to DDT itself, the main breakdown products, DDD en DDE, are in fact almost as persistent as DDT, and if used regularly accumulate in the soil. DDT is practically insoluble in water. DDT accumulates in fish and is therefore a hazard to fish-eating humans and animals. For most animals e.g. snakes, lizards, bees and especially birds DDT is a dangerous pesticide. Many insects which were treated with DDT in the past are now resistant to the pesticide. At present that amounts to some 230 kinds of arthropods, including a number of different kinds of malaria mosquitoes.

Registration: Since the 1960s a large number of countries have prohibited the use of DDT. The main reasons given for this were: the health hazard for humans, the effect on the environment (persistence) and its ineffectivity due to the development of resistence. A few countries still permit the use of DDT for the control of malaria.

DICHLORVOS   insecticide

Trade names: Nogos, Nuvan, Vapona, Dedevap, Benfos, Bibisol, Canogard, Coopervap, Devipan, Dyvos, Erasekt, Mafu, Marvex, Mutox, Nutrax, Phosvit, Roxo.

Properties: Dichlorvos is a colourless to amber liquid, with an aromatic odour, b.p. 35 C/0.05 mm Hg, solubility 10 g/l water. It is stable to heat but is hydrolised by water.

Mode of action: It is a contact and stomach poison, with fumigant and penetrant action.

Uses: It is an insecticide used as a domestic and public health fumigant, especially against flies and mosquitoes; for the protection of stored products at 0.5-1.0 g a.i./100 m^3; for crop protection against sowing and sucking insects at 300-1000g/ha. It is non-phytotoxic and non-persistent. It is used in animal food as an anthelmintic.

Toxicology: WHO-classification: Ib.

Short term: Acute oral LD-50 for rats 56-108 mg/kg. Acute percutaneous LD-50 for rats 75-210 mg/kg. Inhalation LC-50 for rats 15 mg/m^3.

Long term: No long-term effects are observed.

Safe use: For safe use see I.3.3 class Ib. Avoid inhalation of spray mists and contact with the eyes and skin. Do not use in the immediate vicinity of water. Do not consume alcohol before or during spraying. For more precautions and treatment of poisoning see parathion.

Environmental effects: It is highly toxic to honeybees and fish, and toxic to birds.

Registration: Dichlorvos is not prohibited anywhere as far as we know.

DIFLUBENZURON   insecticide

Trade names: Dimilin.

Properties: Diflubenzuron is an off-white to yellow crystalline solid, m.p. about 230 ˚C, solubility 0.1 mg/l water. The solid is stable to sunlight.

Mode of action: It belongs to a group of insecticides, effective as stomach and contact poison, acting by inhibiting chitin synthesis and so interfering with the formation of the cuticle. Hence, all stages of insects that form new cuticles are susceptible to diflubenzuron exposure. It has no systemic activity and does not penetrate the plant tissue, so sucking insects are usually unaffected, forming the basis of selectivity in favour of many insect predators and parasites.

Uses: It is used against: leaf-feeding larvae and leaf-miners in forestry, citrus, field crops including cotton and soyabeans. The larvae of mosquitoes (20-50 g/ha water surface) and flies can be killed by the product.

Toxicology: WHO-classification: III.

Short term: Acute oral toxicity to rats 4500 mg/kg.

Long term: Diflubenzuron is excreted by mammals quite rapidly. No long-term effects were observed.

Safe use: For safe use see I.3.3 class III. Observe the usual precautions, keep out of the reach of children and store away from food and feeding stuff.

Environmental effects: Diflubenzuron breaks down rapidly in the soil (50% breakdown occurring in less than 7 days). It is not very toxic to fish and birds.

Registration: Diflubenzuron is not prohibited anywhere as far as we know.

DIMETHOATE   insecticide

Trade names: Cygon, Daphene, Devigon, Dimetate, Perfekthion, Rebelate, Rogor, Roxion, Trimetion.

Properties: Dimethoate is an organophosphorous compound. Pure dimethoate forms colourless crystals, m.p. 45 ˚C, solubility 25 g/l water. It is relatively stable in aquatic media but hydrolyzed rapidly in alkaline solutions. Under normal storage conditions it may be stored for at least 2 years.

Mode of action: Dimethoate is an indirect cholinesterase inhibitor, see under parathion. It is a systemic insecticide/acaricide and therefore absorbed by the plant. Hence, sucking insects ingest poisoned plantsap. Gnawing insects ingest relatively little poison. Dimethoate is not poisonous itself but is rapidly converted to the actual poison when ingested.

Uses: Dimethoate is used on fruit, maize, grain, soyabeans, vegetables, lucerne and cotton against sucking insects and mites. It is a non-selective pesticide.

Toxicology: WHO-classification: II.

Short term: Acute oral LD-50 for rats 250 mg/kg, lowest dermal LD-50 600 mg/kg. Especially poisonous if taken orally, less so if inhaled or absorbed through the skin. Poisoning can be fatal even at doses of 0.25 gram. Symptoms are the same as for organic phosphorous compounds (see parathion).

Long term: NEL for rats: 1.0-32 mg/kg diet. Dimethoate breaks down and is excreted rapidly (80-90% of the ingested amount within 24 hours). In consequence there is no danger of accumulation. There is still no clear indication about possible negative long-term effects.

Safe use: For safe use see I.3.3 class II. When poisoning occurs, follow the general procedure (I.1). For specific measures when using organic phosphorous compounds see under parathion.

Environmental effects: Dimethoate is hazardous for many organisms (fish, birds, bees). Breakdown in the environment is fairly rapid and there is no risk of accumulation.
Registration: It is not prohibited anywhere.

DIQUAT   herbicide

Trade names: Cleansweep, Reglone, Pathclear.
Properties: Diquat is a dipyridilium-compound. The dibromide monohydrate forms colourless to yellow crystals, decomposing 300 ˚C. Solubility 700 g/l water. Unstable at 9H It may be inactivated by inert clays.
Mode of action: Contact herbicide with an action comparable to that of paraquat.
Uses: Diquat is used in the control of wide-leaved weeds in various crops, against water weeds, to kill off foliage and as a withering media in seed crops, 400-800 g/ha.
Toxicology: WHO-classification: II.
Short term: Acute oral LD50 for rats 231 mg/kg, for cows 30 mg/kg. If inhaled it can cause nose-bleeds. Skin-contact will cause red patches and blisters and can cause nails to fall out.
Long term: NEL for rats: 25 mg/kg diet. Regular exposure can lead to permanent eye injury (cataracts). Diquat is not stored in the body and is excreted within a few days. It is probable that some of the pesticide is broken down by bacteria in the stomach intestinal tubes.
Safe use: For safe use see I.3.3 class II. Working with any commercial preparation without proper protection is always extremely dangerous. Observe the precautions as for paraquat. There are no known restrictions concerning entering a treated field.
Environmental effects: The effects on the environment are very similar to those of paraquat. In the soil diquat is made inactive by absorption to soil particles or by UV rays. Evaporation and rinsing away occurs hardly if at all. If used repeatedly there is a real risk of accumulation in the soil. Some water organisms are very sensitive to diquat. There are no apparent risks for animals which have eaten sprayed plants.
Registration: Diquat is forbidden in many EC-countries because of persistence in the soil.

DISULFOTON   insecticide

Trade names: Disyston, Dithiosystox, Frumin AL, Solvirex, Ekatin-TD, Granulox, Parsolin, Solvigran.
Properties: Disulfoton is an organophosphorous compound. Pure disulfoton is a colourless oil, with a characteristic odour, b.p. 62 C/0.01 mm Hg, solubility at 22 ˚C: 25 mg/l water, readily soluable in most organic solvents. It is stable at storage.
Mode of action: It is a cholinesterase inhibitor with systemic action.
Uses: It is a systemic insecticide and acaricide used mainly for treating seeds and is applied to soil or plants in granule form. It is metabolised in plants, animals and soil.
Toxicology: WHO-classification: Ia.
Short term: Acute oral LD-50 for rats 2.6-12 mg/kg. Acute percutaneous LD-50 for male rats 20 mg/kg. Disulfoton is acutely toxic if swallowed or splashed on the skin.
Long term: In 90 days' feeding trials NEL for rats was 1 mg/kg diet. In mammals it is rapidly eliminated, so no accumulation-effect.
Safe use: For safe use see I.3.3 class Ia. Avoid skin contact when using the granular form. For other precautions see under parathion.
Environmental effects: Moderately toxic to fish, toxic to bees.
Registration: Prohibited in some EEC-countries.

DNOC (dinitro-o-cresol)
insecticide, herbicide, fungicide

Trade names: Antinonnin, Sinox, Nitrador, Selinon, Trifocide, Trifina.
Properties: DNOC is a phenolic compound. Other phenolic compounds are: dinoseb, dinoseb-acetate, pentachlorophenol (see elsewhere in this appendix) and dinoterb. All of these have characteristics very similar to those of DNOC. The technical product forms yellow crystals, explosive when dry, m.p. 85 ˚C, solubility 13 mg/l water.
Mode of action: DNOC acts on fundamental life-processes. The respiratory and metabolic processes in particular are disturbed to such a degree that the very high oxygen consumption causes the reserves in the body to be used rapidly.
Uses: DNOC is toxic to almost all living organisms. It is used as insecticide and acaricide with predomi-

nantly ovicidal action. It is also a contact herbicide with foliar action and it has a secondary fungidal action. It controls post crop emergence of broad leaved weeds in cereals.

Toxicology: WHO-classification: Ib.

Short term: Can be poisonous if swallowed, if vapour is inhaled and, particularly, if absorbed through the skin: DNOC penetrates the skin easily, and especially if the skin is damaged. Even in doses of under 1 gram (ml) of the commercial product, poisoning can be fatal. Symptoms of poisoning are very similar to those of pentachlorophenol (PCP). DNOC can also cause blistering of the skin and can irritate the eyes.

Long term: DNOC is excreted in sweat and urine but very slowly, it can last weeks before it is totally excreted out of the body. If used over a number of days consecutively the danger of poisoning are great. Little is known about the long-term effects. However, it is almost certain that the pesticide causes (temporary?) infertility in males.

Safe use: For safe use see I.3.3 class Ib. If the prescribed safety measures cannot be observed then it is strongly advised not to use DNOC. Protective clothing should be of impermeable plastic or rubber. DNOC should not be used by those with liver, kidney, heart or stomach afflictions or by alcoholics. With DNOC and related pesticides there is a serious risk of fire and explosions! Do not enter treated fields for at least 24 hours after spraying.

Environmental effects: Should DNOC get into the soil, a part will disappear by physio-chemical processes (rinsed away, broken down by UV light) and a part will be broken down by micro-organisms. There is no danger of accumulation. DNOC is highly toxic to all sorts of living creatures both above and below ground, this also applies to birds, bees and mammals. Hence if possible do NOT use DNOC!

Registration: DNOC is not prohibited in any country.

ENDOSULFAN   insecticide

Trade names: Beosit, Cyclodan, Malix, Thiodan, Thifor.

Properties: Endosulfan is an organochlorine, it is a brown crystalline solid with an odour of sulphur dioxide; m.p. 70-100 °C, solubility 0.32 mg/l water.

It is stable to sunlight, unstable in alkaline media and subject to slow hydrolysis.

Mode of action: Wide range non-systemic contact- and stomach insecticide. It acts via overactivation of the nervous system.

Uses: Non-selective pesticide used to control insects, especially beetles, caterpillars, and aphids, also used against mites in fruit trees, vegetables, rice and other food crops. Also used on cotton, tabacco and tea.

Toxicology: WHO-classification: II.

Short term: LD-50 for rats is about 80 mg/kg. Endosulfan is moderately toxic to mammals including humans. Poisoning can occur if taken orally, absorbed through the skin or inhaled. Symptoms of poisoning are, among others: headaches, loss of appetite, temporary deafness, see also organochlorines (under DDT). Poisoning is often caused by eating contaminated food. A protein-deficient diet heightens the effects of poisoning.

Long term: Endosulfan is decomposed and excreted quite rapidly in the faeces and urine. There is no real risk of accumulation in the human body. There are no definite indications of mutagenity, carcinogenity or negative effects on fertility.

Safe use: For safe use see I.3.3 class II. See general precautions, treatment in the event of poisoning: as for other organochlorines (see under DDT). No specific antidote is known. Endosulfan must not be used by liver or kidney patients.

Unless sufficiently protected, treated fields should not be entered for at least two days after spraying.

Environmental effects: In comparison to other organochlorines endosulfan breaks down fairly rapidly (residual activity in soil a few months). Breakdown in humid soils is quicker than in dry ones. The pesticide is very poisonous to fish and other aquatic animals; a dose of 100 gram per hectare will cause the disappearance of all fish. Therefore, keep endosulfan well away from water (especially relevant for wet rice cultivation). Treated crop remains must not be used as cattle fodder. Higly toxic to bees and birds.

Registration: Forbidden in many EC-countries because of fish toxicity. In other countries use is severely restricted.

EPN
insecticide

Properties: Pure EPN is a light yellow crystalline powder, m.p. 36 °C, practically insoluble in water. The technical grade is a dark-amber-coloured liquid. Incompatible with alkaline pesticides.
Uses: It is a non-systemic insecticide and acaride with contact and stomach action. It is effective at 0.5-1.0 kg a.i./ha against a wide range of lepidopterous larvae, especially bollworms and Alabama argilacea on cotton, rice stem borers and other leaf-eating larvae on fruit and vegetables. Usually non-phytotoxic.
Toxicology: WHO-classification: Ia.
Short term: Acute oral LD-50 for female rats 14 mg/kg. Acute dermal LD50 230 mg/kg. EPN is very toxic to mammals, especially when swallowed.
Long term: Only effect observed was retarded growth at max. 450 mg/kg diet for male rats.
Safe use: For safe use see I.3.3 class Ia. Strictly observe general precautions for use of pesticides.
Environmental effects: Moderately toxic to birds and very toxic to bees and fish.
Registration: EPN is restricted or forbidden in many EEC-countries.

ETHYLENE DIBROMIDE
insecticide

Trade names: Bromofume, Dowfume 85, Agrifume, Celmide, Edabrom, Nemafume, Pestmaster.
Properties: It is a colourless liquid, b.p. 131.5 °C, m.p. 9.3 °C, solubility 4.3 g/l water.
Uses: It is an insecticidal fumigant used against pests of stored products; for the treatment of fruit and vegetables; for the spot treatment of flour mills; for soil treatment against certain insects and nematodes. Planting must be delayed until 8 days after soil treatment because of its phytotoxicity.
Toxicology: WHO-classification: II.
Short term: Acute oral LD50 for female rats 146 mg/kg. Dermal applications, if confined, will cause severe burning of the skin.
Long term: Ethylene dibromide is carcinogenic.
Safe use: For safe use see I.3.3 class II. No specific antidote known. Symptomatic treatment: wash exposed skin with soap and water. Rinse exposed eyes for more than 15 minutes and get medical assistanse. After inhalation, remove patient to fresh air.

If swallowed, induce vomiting with warm salt solution. Call a doctor.
Registration: Forbidden in many ECC-countries because of its carcinogenicity.

FENITROTHION
insecticide

Trade names: Accothion, Cytel, Cyfen, Folithion, Sumithion.
Properties: Technical grade fenitrothion is a yellow-brown liquid, solubility 14 mg/l water. It is an organophosphorous insecticide.
Mode of action: It has contact, stomach, and respiratory action; a cholinesterase inhibitor.
Uses: Fenitrothion is a potent contact insecticide, effective against a wide range of pests i.e. penetrating, chewing and sucking insect pests (coffee leafminers, locusts, rice stem borers, wheat bugs). It is also effective against household insects. It is effective as vector control agent for malaria and has been approved as an insecticide for locust control.
Toxicology: WHO-classification: II.
Short term: Acute LD-50 for female rats 800 mg/kg. Relatively low toxicity to mammals. It is harmful by inhalation and if swallowed and in contact with skin. It is irritating to eyes.
Long term: In feeding trials NEL for rats was 5 mg/kg diet.
Safe use: For safe use see I.3.3 class II. It is harmful to livestock so keep all livestock out of treated areas for at least 7 days. Do not consume crops until 2 weeks after using fenitrothion. Do not re-use the container for any other purpose and if you feel unwell seek medical advise.
Environmental effects: Harmful to birds and mammals. Dangerous to bees. Harmful to fish.
Registration: Restricted in many EEC-countries.

GLYPHOSATE
herbicide

Trade names: Roundup, Tumbleweed.
Properties: Pure glyphosate forms colourless crystals, m.p. 200 °C, solubility 12 g/l water, it is strongly absorbed in soil in which decomposition is mainly by microbial activity.
Mode of action: The pesticide is absorbed through the parts of the plant rising above the surface and has the effect of disturbing the plant's me-

tabolism. Incorrect use can cause serious crop damage.

Uses: Glyphosate is a herbicide used in many crops. It is very effective on deep-rooted perennial species, and annual and biannual species of grasses, sedges and broad-leaved weeds. Dosage 0.34-1,12 kg/ha for annual, 1.68-2,24 for perennial species.

Toxicology: WHO-classification for toxicity: III.

Short term: Acute oral LD-50 for rats 4050 mg/kg. Glyphosate is a lightly toxic pesticide which can have an irritating effect on the eyes.

Long term: As far as is known it has no long term damaging effects. It does not accumulate in the body.

Safe use: For safe use see I.3.3 class III. There are no restrictions as far as entry into treated fields is concerned.

Environmental effects: Residual activity in the soil is about a month, therefore it has no cumulative effect. As glyphosphate is a 'total-herbicide' care must be taken that useful crops do not suffer damage

Registration: Glyphosphate is nowhere prohibited. On the contrary: within the EEC and in other countries too, its use is being increased and its application admitted in ever more situations.

HEPTACHLOR   insecticide

Trade names: Velsicol 104.

Properties: Heptachlor is an organochlorine. The technical grade of heptachlor contains a mixture of compounds (72% heptachlor) and is a waxy solid; m.p. 46-74 °C. It is stable to light, moisture and air °C. It is not readily dehydrochlorinated but is susceptible to epoxidation.

Mode of action: As for other organochlorines.

Uses: Used to control all sorts of insects in grain, maize, banana, oil containing seeds, vegetables, sugar-cane, nuts. It is applied directly to foliage and used to treat seeds and to disinfect soil. Also used to control household insects, pests of man and domestic animals.

Toxicology: WHO-classification II.

Short term: Acute oral LD-50 for rats 147-220 mg/kg.
Oral intake and inhalation is very damaging. Symptoms are the same as for other organochlorines. In

serious cases victim can suffer convulsions and go into coma.

Long term: NEL for rats 5 mg/kg diet, for dogs 1 mg/kg diet. The pesticide is converted in the body into heptachlor epoxide which is 2-4 times as poisonous as heptachlor itself. This conversion product easily accumulates in the fat-tissues of the body. Regular exposure to the pesticide gives great risks of liver afflictions. The pesticide is also suspected to be carcinogenic.

Safe use: There is little point in advising safety measures for a pesticide on the long term so toxic as heptachlor. The only possible advice must be: DO NOT USE.

Environmental effects: Very toxic to aquatic organisms. Some phytoplankton species are killed at concentrations as low as 1 g per l water (1:109). Shrimps and various types of fish are even many times more sensitive. Alongside this, the chemical is very persistent in the environment, and it has a strong cumulative effect in the food chains.
In granular form heptachlor is easily carried in the wind from land to water and consequently damage water life. In areas where a lot of heptachlor is used a serious drop in the bird population has been observed. The pesticide is also very toxic to bees.

Registration: Totally or partly prohibited in many countries.

HEXACHLOROBENZENE   fungicide

Trade names: HCB, Antie-carie, Bunt-no-more, HCB, Hexa-CB, Sanocide.

Properties: Pure hexachlorobenzene forms colourless crystals, m.p. 226 °C, practically insoluable in water.

Mode of action: Possibly by fumigant action on fungal spores.

Uses: It is a selective fungicide.

Toxicology: WHO-classification: III.

Short term: It is considered relatively non-toxic and non-hazardous in handling. Acute oral LD-50 for rats 10,000 mg/kg, it may cause a slight irritation to the skin.

Long term: We could not find data on long term effects.

Safe use: For safe use see I.3.3 class III. Do not inhale the spray mist and keep it away from children and domestic animals.

Registration: The use of HCB as well as the presence of residues in food is forbidden in many countries.

LINDANE   insecticide

Trade names: BHC, HCH, Gammexane.
Properties: Lindane is an organochlorine, which consists of a mixture of isomeres of hexa- chloro-hexane, the gamma-isomere content of which is at least 99%. Also known as gamma- HCH, or gamma-BHC. Due to the mixture it has no precise physical properties. M.p. about 112 ˚C, stable œ ˚C, and to light. It is decomposed by alkali. Note: HCH is a mixture of isomeres with only 10-20% of the hexachlorhexane gamma-isomere. The remainder is non-active and is nothing more than chemical waste. For this reason its use is not recommended.
Mode of action: It acts as a stomach poison, by contact, and has some fumigant action. De gamma-isomere is the only active pesticide, it has the same effect as all organochlorines: overstimulation of the central nervous system.
Uses: Broad spectrum insecticide used against all types of pests in agriculture and horticulture and especially for treating soil and disinfecting seeds. Is also sometimes used as bait in the control of rodents, and to treat cattle against external parasites. In the past its was commonly used to control vectors of diseases such as malaria.
Toxicology: WHO-classification: II.
Short term: Lowest LD-50 for rats 88 mg/kg. Acute percutaneous LD-50 for rats 900 mg/kg. The pesticide can be inhaled, swallowed and absorbed through the skin. Lindane is one of the more toxic of the organochlorines; particularly in case of malnutrition and protein deficiencies. The pesticide can irritate the eyes, the skin and the repiratory organs. HCH can cause severe irritations. Symptoms are the same as for other organochlorines.
Long term: NEL for rats 25 mg/kg diet. Lindane is partly accumulated in the body, mainly in the fat-tissues but also in the liver and in the brain. This accumulation level is greater for less pure HCH products. Breast-feeding mothers can pass the pesticide on to their babies.
Research has indicated that lindane/HCH can cause cancer of the liver and thyroid gland. Lindane/HCH

is also toxic to embryos, and impedes the bodies immunity system.
Safe use: For safe use see I.3.3 class II. For treatment of poisoning see organochlorines (under DDT).
Environmental effects: Lindane is an accumulative pesticide. HCH has a residual activity period in the soil of a number of years. It leaves the soil and water through evapouration processes but also through degrading by ultra-violet light. Soil treated with lindaan may not be used for cultivating food-crops for between 1 and 3 years. Lindane is poisonous to fish, especially for salt water species. Also poisonous to birds because of the risks caused by the cumulative effect for weak egg-shells.
Registration: The BHC/HCH-products which accumulate in the environment more than lindane are totally prohibited in EEC countries and in the US. Lindane itself is admitted in most countries. In the Netherlands it will be prohibited as from 1991.

MALATHION   insecticide

Trade names: Carbofos, Cythion, Maldison, Malathon, Mercaptation.
Properties: Malathion is an organophosphorous compound. 95% pure malathion is a clear amber liquid, m.p. 3 ˚C, solubility 145 mg/l water.
Mode of action: It is a non-systemic insecticide and acaricide. The pesticide is transposed into malaxon, the actual active material.
Uses: Malathion is used in many field and garden crops to control insects and for cattle against animal ectoparasites, in storage spaces, and against malarial mosquitoes and their larvae.
Toxicology: WHO-classification: III.
Short term: LD-50 for rats 2800 mg/kg. Acute percutaneous LD50 for rabbits 4100 mg/kg. In comparison with paratahion, malthion is far less dangerous. Fatal poisoning can occur at an intake of 4-5 grams. Symptoms of poisoning are the same as for other organophosphorous compounds. Malathion causes hardly any accidents with fatal consequences. The main danger is that the product can contain pollutants such as isomalathion. This transposed material delays natural breakdown of the poison in the body which can cause the toxic effect of malthion to increase considerably.
Long term: NEL for rats 100 mg/kg diet. Malathion does not accumulate in the body. As far as is

known, it has no cell-modification properties, is not carcinogenic or responsible for foetal abnormalities. It must be noted here that in spite of extensive use of this insecticide, these aspects have not yet been fully researched.

Safe use: For safe use see I.3.3 class II. Due to the possibility of technical pollution (isomalathion) safety regulations should be observed more stringently than is normally the case for chemicals classified in group III. For help in cases of poisoning look for organophosphorous compounds under parathion.

If not wearing protective clothing, avoid entering fields for one day after treatment.

Environmental effects: Malathion breaks down rapidly in the environment, half-life approximately one month, mainly due to fungi and bacteria. The chemical is toxic to fish and bees and moderately toxic to birds.

Registration: Malathion is practically nowhere prohibited.

MCPA en MCPP   herbicide

Trade names: MCPA: Agroxone, Hedonal M, Isocornox, Weedone. MCPP: Mecoprop.

Properties: MCPA and MCPP are phenoxy acetic acids. Pure MCPP forms colourless crystals, m.p. 90 °C. Solubility 620 mg/l water. It is stable to heat, resistant to hydrolysis.

Mode of action: Similar to 2,4-D.

Uses: MCPP is used after emergence to control weeds, at 1.5-2.7 kg/ha. It is mainly used in combination with other herbicides to extend the range of weeds controlled.

Toxicology: MCPA is classified in class III.

Short term: MCPP has an acute oral LD-50 for rats of 930 mg/kg. A fairly poisonous pesticide particularly if taken by mouth or through the skin. Symptoms of poisoning: unconsciousness, cramps similar to epilepsy, heavy sweating. If MCPA has been in contact with the skin, blisters and minor burns develop. A fatal incident has been recorded of someone who drank 60 ml of the commercial preparation.

Long term: It is suspected that MCPA is able to cause modifications in cell material but it is not clear if the chemical is carcinogenic. In rats and mice it effects the sperm-cell production and after

high dosages abnormal embryos develop. Considering its world-wide use, little is known about the chemical.

Much the same about MCPP although its acute toxicity is somewhat less than that of MCPA.

Safe use: For safe use see I.3.3 class III. See further 2.4-D.

Environmental effects: MCPA and MCPP ecompose relatively quickly in the environment mainly due to various species of bacteria. The half-life ranges from several weeks to months. In soil that has received previous treatment, breakdown will occur more quickly because the bacterial flora has already adapted.

MCPA is fairly poisonous to birds. Birds have been known to die after flying through drifting clouds of spray. MCPA is fairly safe for fishes.

Registration: MCPA is forbidden in most EC-countries and replaced by MCPP.

MONOCROTOPHOS   insecticide

Trade names: Nuvacron, Azodrin.

Properties: Organophosphorous compound. It forms colourless crystals, m.p. 54 °C, solubility 1 kg/l water.

Mode of action: It is a fast-action insecticide with both systemic and contact action.

Uses: Against above-ground insects in crops such as rice, sugar-cane, citrus fruit, tobacco, potatoes. Nonselective. Dosage 230-1000 g/ha. It persists for 7-14 days.

Toxicology: WHO-classification: Ib.

Short term: Acute LD-50 for rats 14 mg/kg. Acute percutaneous LD-50 for rats 336 mg/kg. Monocrotophos has an especially high acute toxicity when orally taken, but inhalation and skin-contact are also hazardous. Symptoms of poisoning are as for other organophosphorous compounds (see under parathion).

Safe use: For safe use see I.3.3 class Ib. Observe general precautions and as named for organophosphorous compounds (see parathion). Do not enter sprayed area for several days after treatment.

Environmental effects: Monocrotophos rapidly breaks down in the soil, its half-life is approximately one month. There is no danger of accumulation in the soil. The preparation is very toxic to birds, fish, crabs, shrimps and bees among others. It must

be stressed that pollution of water must be avoided.
Registration: No application has been made in
the EEC for permission to use monocrotophos. In
India there are strict limitations to its use.

NITROFEN   herbicide

Trade names: Trizilin, Tok E-25, Tokkorn, NIP.
Properties: It is a crystalline solid, m.p. 71 °C, solubility at 22 °C about 1 mg/l water.
Mode of action: Selective contact-herbicide, predominantly in pre-emergence application. Requires
light in order to work, therefore it should not be
worked into the soil.
Uses: Nitrofen is a selective herbicide, toxic to a
number of of broad-leaved and grassy weeds, and effective when left as a thin layer on the soil surface.
It is used on cereals pre-emergence at 2 kg a.i./ha.
Toxicology: WHO-classification: III.
Short term: Acute oral LD-50 for rats about 700 mg
a.i./kg. Neither the a.i. nor the formulations caused
irritation to skin of rabbits or visible toxic effects.
Long term: Preliminary investigations carried out
under normal experimental conditions indicate that
there is no danger of chronic damage. Highest dose
without activity in 90-day tests on rats: 3.45 mg/kg
body wt./day. May be carcinogenic.
Safe use: For safe use see I.3.3 class III. Women of
child-bearing age and children should not be exposed to the product. Obey general precautions. Prevent drifting during use.
Environmental effects: Toxic to fish. Not toxic
to bees. The pesticide is degraded in the soil by
microbes. Duration of residual activity in soil ca. 6
weeks (after 10 kg emulsifiable concentrate/ha).
Registration: Forbidden in EC because of supposed carcinogenicity.

PARAQUAT   herbicide

Trade names: Dextrone X, Esgram, Gramoxone;
in various mixtures: Cleansweep, Dexuron, Gra-
monol, Gramuron, Para-col, Pathclear, Tota-col.
Properties: Dipyridilium-compound. There is a dichlorid and a salt-form. The dichlorid forms colourless crystals decomposing at 300 °C. It is very
soluble in water. The salts are stable in neutral and
acid media but are oxidised under alkaline condi-

tions. They are inactivated by inert clays.
Mode of action: Green plant parts which come
into contact with paraquat burn, an effect which is
stronger under light conditions. Paraquat prevents
photosynthesis in leaves. The plant probably suffers
most from hydrogen peroxide, toxic to the plant,
which is formed under the influence of paraquat.
Uses: Paraquat is a contact herbicide against both
broad-leaved and grassy weeds in all kind of cultures. Also commongly used against water weeds
and further as a defoliant in a number of crops: potatoes, maize, soyabeans, cotton. Uses include inter-
row weed control in vegetable crops.
(560-1120 g/ha) weed control in plantation crops
(280-560 g/ha).
Toxicology: WHO-classification: Ib.
Short term: Lowest acute oral LD-50 for rats is 150
mg a.i./kg, for dogs 25 mg a.i./kg. Paraquat is a pesticide with a specially high acute toxicity. Intake of
a very small quantity i.e. a teaspoon full, of the
commercial dilution is fatal. It irritates mouth,
throat and stomach; liver and kidneys are also seriously affected in case of poisoning. The most serious effect is hardening of the lung-tissues, causing
decreasing effectiveness of lung function. Medical
treatment is not possible for this. When paraquat is
inhaled, the muceous membranes are affected and
nose-bleeding can result. Skin-contact with paraquat is equally poisonous: red spots or blisters appear on the skin, nails fall out and eye injuries
become apparent. These symptoms usually disappear after a fairly short time. About half of all cases
of paraquat poisoning result in death. It is estimated that this chemical causes about 100 deaths
in Papua-New Guinea annually.
Long term: NEL for rats 170 mg/kg diet, for dogs
34 mg/kg diet. Paraquat is suspected of being carcinogenic and affects gene material. In rats offspring
irregularities occurred while at the same time fertility diminished. The long-term effects of paraquat
have been inadequately researched. Paraquat does
not accumulate in the body; it is rapidly excreted.
Safe use: For safe use see I.3.3 class Ib. The use of
paraquat should only be permitted if the users are
well protected: impermeable clothing, face-masks,
head-covering, gloves, boots etc. This also applies to
working with the diluted product.
If poisoning occurs, immediately induce vomiting
(although from experience this does not help
much), make the victim eat clay-soil which is the

only means to neutralize paraquat to some extent. Hospitalization is absolutely necessary in all cases. Plenty of water should be drunk. Skin contact areas should be rinsed thoroughly with water.

Environmental effects: Paraquat is absorbed easily by soil-particles and inactivated. Decomposition by soil organisms hardly or never takes place. Generally the absorption capacity of the soil is adequate to take in several hundred times the quantities of paraquat normally used. However, in soil deficient in clay minerals and with a high proportion of humus, the absorption capacity can quickly be exceeded. In the long term, paraquat can be a threat to the arable quality of the field.

Because of the rapid absorption by soil particles, there is little danger to animals on land or in the water, but there is indeed danger to animals who consume the plants shortly after they have been sprayed. Toxic to bees.

Registration: Forbidden in many EEC-countries because of persistence in soil. In Malaysia, paraquat must contain an evil-smelling chemical to prevent it being drunk by mistake, while in Papua-New Guinea it must contain an emetic.

PARATHION   insecticide

Parathion is an organophosphorous compound. First some general points about these compounds will be mentioned, then some further attributes of parathion in particular.

Organophosphorous compounds.

Mode of action. Organophosphorous compounds are anticholinesterases, usually powerful poisons for both insects and mammals. The primary effects are attributable in part or entirely to the inhibition of an enzyme, acetylcholinesterase, in nervous tissues. It causes disturbances in glands, muscles and in parts of the brain.

There are two groups of organophosphorous pesticides: direct and indirect inhibitors. After exposure to direct inhibitors, symptoms usually appear fairly quickly. After exposure to indirect inhibitors, which first have to be transformed in the body to an active compound, a more delayed and prolonged action appears. When symptoms appear, they may develop further to **cause a critical illness.**

Symptoms of poisoning. The first symptoms of poisoning are: nausea, headache, a feeling of weakness,

mental confusion and muscular non-coordination. Vomiting, pain and diarrhoea, excessive sweating and dribbling, muscular twitchings. In severe cases paralysis soon follows, then unconsciousness with convulsions may precede respiratory failure, the usual cause of death.

Emergency treatment. See under general advice for cases of poisoning. Make the patient vomit. Atropine is the most common antidote used against organophosphorous poisoning. The patient should be injected by a doctor with 2-10 mg atropine and then every 10-20 minutes continued with 2-4 mg, until the patient is fully atropinized, up to 100 mg or more within 24 hours. An overdose of atropine is rarely serious, an underdose may be fatal. Oximes (e.g. pralidoxime) are other antidotes, to be applied by a doctor.

Important trade names of parathion: Bladan, Folidol, Fosferno, Niran, Thiophos.

Properties: Technical parathion is a brown liquid with garlic odour, b.p. 157-162 °C, solubility 24 mg/l water. It is rapidly hydrolized in alkaline media.

Mode of action: It is a non-systemic contact and stomach-acting insecticide and acaricide with some fumigant action. parathion attacks the central nervous system. In the body, parathion is converted to paraoxon, the actual active compound.

Uses: This insecticide is used as an insecticide with a broad application for soil and leaf treatment in many crops: fruit-trees, cotton, maize, cereals, leaf- and root-vegetables, tomatoes, tobacco etc.; against mosquito larvae in rice, agaisnt soil-insects in meadows.

Toxicology: WHO-classification: Ia.

Short term: Acute oral LD-50 for rats 3.6 mg/kg. Lowest acute percutaneous LD-50 for rats 7 mg/kg. Parathion is extremely toxic and even more in the event of a protein-deficient diet. The pesticide can equally easily be absorbed into the body via the mouth or through the skin.

Of all the agricultural toxic pesticides parathion causes the greatest number of fatal poisonings. In Central America, 80% of poisonings through control pesticides are caused by parathion. Accidents occur when the pesticide is being used, but also through eating contaminated food, i.e. by transporting sugar in waggons which had earlier transported parathion, or by using old packages for the storage of flour etc. Symptoms of poisoning are the same as for other organophosphorous compounds (see above).

Long term: No accumulation in the body; no further effects known.

Safe use: For safe use see I.3.3 class Ia. If protective clothing is not available, the use of parathion must be strongly advised against. Old packaging material must not be used for any other purpose. Without proper protection, people must be kept away from sprayed crops for at least several days.

Environmental effects: Parathion is an extremely toxic pesticide for all kinds of organisms: fish, birds, bees, domestic animals and cattle. Breakdown in the soil occurs quickly with the period of residual activity from several weeks to a few months. There is no danger of accumulation.

Registration: Permitted in EEC countries, however forbidden in Japan and South Africa.

PENTACHLOROPHENOL
                    insecticide, herbicide, fungicide, etc.

Trade names: Dowicide EC7, Dowicide G, PCP, Penta, Santobrite.

Properties: Pentachlorophenol is a substituted phenolical compound (like DNOC). Technical grade of PCP is dark grey, m.p. 187-189 °C, with a phenolic odour, it is a weak acid, solubility 330 g/l water.

Mode of action: Disturbs the metabolism in nearly all living organisms and also affects the nervous system.

Uses: To cause withering in cotton, as herbicide in various crops, as a fungicide especially to preserve timber, as insecticide and fungicide in the treatment of seeds, to control snails etc.

Toxicology: WHO-classification: Ib.

Short term: Lowest acute oral LD50 for rats 27 mg/kg. The pesticide is mainly taken into the body through the mouth, through breathing and also largely through the skin. The lethal dose for humans is about 2 gram. The symptoms of poisoning are: initially a feeling of fitness, fitter than normal. Increase in body-temperature. The result of poisoning can lead to the victim himself increasing the amount of spraying and working a longer day. After several days other symptoms become obvious: much sweating, great thirst, tiredness, loss of weight which in hot weather will not be noticeable at the start. Other indications are: diarrhoea, stomach ache, sickness, vomiting and dizziness. The toxicity is considerably increased to the degree by which the general temperature rises. Another effect of PCP is developing of white blisters on the skin and irritation in the eyes. There is further a chance of damage to the liver and kidneys. Inhaling vapours or mists of PCP can cause breathlessness, followed in serious cases by a state of coma with fatal outcome.

Long term: There is a slight accumulation of the pesticide in the body. There are strong indications that PCP causes embryonic abnormalities, and possibly the pesticide is also carcinogenic. This is not entirely clear as it may be partly caused by the many contaminations which can occur with PCP preparations, amoung them certain dioxines.

Safe use: For safe use see I.3.3 class Ib. In poisoning cases, quick action is essential. When taken orally: rinse the mouth, drink water or milk preferably with charcoal tablets or powder. Thereafter stomach rinse or emetic. The patient must be kept cool and possibly artificial respiration must be applied. If contact has been with the skin, rinse well with much water for at least 15 minutes. After this cleanse the skin well with water and soap. A specific anitdote is not known.

Environmental effects: The pesticide accumulates particulary in water organisms. Therefore it cannot be used against water snails.

Registration: PCP is not forbidden anywhere. In the US the pesticide has been placed on the list of suspicious pesticides because of its possible effect on embryos.

PYRETHRINES en SYNTHETIC
PYRETHROIDS   insecticides

Production: Pyrethrines are pesticides isolated from the flower-heads of the pyrethrum flower (Chrysantemum cinerariaefolium). The insecticidal effect of this extract has already been known since 1800. The present centre of production is Kenya from where the harvested flower heads are exported to China for further processing into a product useable in practice. Because natural pyrethrum products degrade rapidly, particularly through the effect of sunlight, they are suitable for domestic use but less so for agricultural application. The more stable synthetic pyrethroids, among them permethrin, fenvalarate, bioallethrin, cypermethrin and tetramethrin, are however more suitable for this.

Properties: Most commercial extracts contain 20-25% pyrethrins and are pale yellow, the plant waxes and pigments having been removed. They are unstable in light and are rapidly hydrolyzed by alkali with loss of insecticidal properties.

Mode of action: Pyrethrins are potent, non-systemic contact insecticides. In practice they have two important effects:

- A knock-down effect where contact causes rapid paralysis. To prevent the creatures scrambling up again and escaping, a synergist (strengthener) is usually added which stops the rapid decomposition of the pesticide inside the insect's body. This kills the insects at a later stage.
- A repellent effect at low concentrations. This is used in the well known midge-spirals which are allowed to smoulder in the evenings to keep stinging insects away.

Uses: Pyrethrines are used against midges, mosquitos, flies, cockroaches, body-lice and other isects. They are much in use on recently picked fruit, cereals, animal foodstuffs and similar items, and also as a repellent on the outside of food packages.

Toxicology: WHO-classification: II.

Short term: Lowest acute oral LD-50 for rats 584 mg/kg. Acute percutaneous LD-50 1500 mg/kg. The various pyrethroids have a different LD-50. Pyrethrines are poisonous to insects including bees, but hardly for warm-blooded animals. The lethal dose for humans is in the region of 100 gram, and therefore there is little risk of poisoning. The most important problem appears to be some appearance of skin irritation or asthma in persons susceptible to this (asthma and hay-fever sufferers). Some pyrethroids have a low LD-50 but are applied in very low concentrations, thus not presenting much danger.

Long term: Pyrethrines and pyrethroids are rapidly degraded in and excreted from the body. Long term effects have not yet been fully researched.

N.B. Usually piperonyl-butoxide is added as a synergist to pyrethrine. There is some uncertainty about the effect on health of this chemical; it may be carcinogenic; however the acute toxicity is small.

Safe use: For safe use see I.3.3 class II. Mask and gloves should be used to avoid skin contact. Should not be used by hay-fever patients. There is no problem regarding entry to treated fields. If poisoning occurs: induce vomiting and rinse stomach with 5% sal volatile.

Environmental effects: Natural pyrethroids degrade rapidly in sunlight in a few hours; in dimmer light it can take days to weeks, depending on conditions such as humidity. Toxicity to warm-blooded animals is relatively small, for cold-blooded ones relatively great: many fishes are affected. The useful quality of synthetic pyrethroids when spraying, that they are more persistent, is a disadvantage for the environment where they can still be found months after their application.

Registration: Permitted everywhere.

2,4,5-T   herbicide

Trade names: Brush killer, Brush Rhap, Transamine, Trioxone, Weedone.

Properties: Chemical group: phenoxy acetic acids. Technical grade 2,4,5-T forms colourless crystals, m.p. 153-156C. Solubility 150 mg/l water. It is stable in aquateous solutions at pH 5-9.

Mode of action: It is absorbed through roots, foliage and bark. Effect as 2,4-D.

Uses: As a post-emergence herbicide, often in mixtures with, for example, 2,4-D, in corn and other crops against broad-leaved weeds.

Toxicology: 2,4,5-T has become infamous through its use as a defoliant during the Vietnam war. The innumerable after-effects (embryonic abnormalities, abortions) from the immediate post-war period can with some certainty be traced back to the presence of a dioxine, TCDD, formed as a by-product during production of 2,4,5-T as well as of some other control pesticides. Dioxine can probably be considered the most poisonous chemical presently known. It produces a serious form of acne (pustules/chloracne) from which recovery is difficult. It further causes impotence, liver damage, injury to the nervous system and the already mentioned effects on offspring. The natural resistance mechanism of the body also greatly suffers through the presence of dioxine.

Modern production methods have now made it possible to lower the dioxine contents of 2,4,5-T considerably; nevertheless it is still not possible to produce the herbicide entirely without this pollutant.

WHO-classification: II.

Short term: Acute oral LD-50 for rats: 300-1700 mg/kg. Acute percutaneous LD-50 for rats 5000 mg/kg. The toxicity picture of 2,4,5-T is noticeably vague, to some extent because research results have not really indicated to what extent the presence of TCDD has been taken into account. The pesticide can enter the body through the mouth and respiratory passages, to a lesser extent through the skin. An intake of 3-4 gram can arouse symptoms of poisoning. These have not been well described, but can be expected to be similar to the effects of 2,4-D.

Long term: NEL for rats 30 mg/kg. The pesticide does not accumulate in the body. Other long term effects are not clear; possibly commerical products, i.e. those containing dioxine, will have an effect on embryos (death, malformation).

Safe use: For safe use see I.3.3 class II. As it is not clear to what extent the commercially available products are weak in dioxine, we advise against the use of 2,4,5-T while this uncertainty persists. A better alternative is 2,4-D or tryclopyr. Pregnant women should keep far away from fields sprayed with 2,4,5-T.

Environmental effects: 2,4,5-T is very poisonous to fish, birds and bees can withstand it better. The pesticide does not accumulate in the food-chains but is rapidly excreted. Micro-organisms in the environment play an important role in the decomposition of this pesticide.

Registration: Because of the presence of dioxine, 2,4,5-T is prohibited in many countries.

TRIAZOPHOS   insecticide

Trade names: Hostathion.

Properties: It is an organophosphorous compound. Pure triazophos is a yellowish oil, m.p. 2-5 °C. Solubility at 20 °C 39 mg/kg water.

Uses: It is a broad spectrum insecticide and acaricide with some nematological properties. It controls aphids on cereals at 320-600 g a.i./ ha, on fruit at 75-125 g a.i./100 l. It is also used to control Lepidoptera on fruit and vegetables. When incorporated at 1-2 kg a.i./ha in the soil prior to planting, it controls Agrotis spp. and other cutworms. It can penetrate plant tissue, but has no systemic activity.

Mode of action: Contact and stomach action. Cholinesterase inhibitor.

Toxicology: WHO-classification: Ib.

Short term: Acute oral LD-50 for rats about 60 mg/kg. Acute dermal toxicity to rats 1100 mg/kg.

Long term: Inhibition of blood serum cholinesterase was the only effect noted in the long term.

Safe use: For safe use see I.3.3 class Ib. Avoid skin contact and inhalation of spray mists. Precautions are those commonly employed with organophosphorus compounds (see parathion). Antidote: atropine plus toxogonin.

Environmental effects: LC-50 for carp ca. 1 mg/l. Toxic to bees.

TRIFLURALIN   herbicide

Trade names: Elancolan, Treflan, Triflurex.

Properties: Technical trifluralin is an orange crystalline solid, m.p. 49 °C, solubility mg/l water. It is quite stable, though susceptible to decomposition by UV-radiation.

Mode of action: Trifluralin inhibits the cell-division of plants after absorption by parts beneath the soil.

Uses: It is a pre-emergence herbicide mainly against grassy weeds, but can also be used against broadleaved in cotton, soya, sugar-beets, vegetables. In view of its great volatility, the pesticide must be worked well into the soil. Normal dose: 500-1000 g/ha.

Toxicology: WHO-classification: III.

Short term: LD-50 for rats 10,000 mg/kg, for mice 500 mg/kg. The acute toxicity of trifluralin is small. On the basis of the rat-data, a human being would have to swallow half a kilo to be in acute danger of death.

Long term: The pesticide does not accumulate in the body. There is some suspicion about the long-term effects of the pesticide because on the basis of experiments with animals there is evidence of mutations and also of harmful effects on offspring.

Safe use: For safe use see I.3.3 class III. Spray wearing protective clothing which should be washed carefully after use. After contact with skin or eyes, immediately wash with much water.

Environmental effects: Degrades under the influence of light and, on the surface, through volatility. In the soil there is residual activity of a few months. Little danger for birds and fish.

Registration: As far we know, not forbidden anywhere. In the US it has been placed on the list of suspicious pesticides.

ZINEB, MANEB and MANCOZEB
fungicides

Trade names: Zineb: Dithane-Z-78, Lonacol, Linate, Tiezene.
Maneb: Dithane M-22, Manzate.
Mancozeb: Dithane M-45, Manzate 200, Manzeb, Sandozebe.
Properties: These are all dithiocarbamates.
Zineb is a light-coloured powder, it decomposes before melting, solubility 10 mg/l water, somewhat unstable to light, heat and moisture.
Maneb is a yellow crystalline solid, it decomposes on prolongued exposure to air or moisture, or rapidly on contact with acids.
Mancozeb is a complex of zinc and maneb, it is a greyish-yellow powder, which also decomposes before melting, it is practically insoluble in water and is stable under normal storage conditions.
Mode of action: The effect is mainly growth-inhibiting.
Uses: Applied as a fungicide on a number of crops, but is ineffective against fungi which have penetrated the plant. Normal dose of Mancozeb: 1.4-1.9 kg/ha.

Toxicology: WHO-classificatin: III.
Short term: For Zineb and maneb oral LD-50s on rats are respectively 5200 and 8000 mg/kg. The acute toxicity of these pesticides is very small. They do have an irritating effect on the eyes (tears, irritation), on the air-passages (bronchitis type symptoms), the skin (irritation) and the intestinal tupe (nausea, vomiting and diarrhoea). Toxic effects increase in combination with alcohol.
Long term: NEL for rats 250 mg/kg diet. The pesticides do not accumulate in the body, but are suspected to have carcinogenic and mutant effect.
Safe use: For safe use see I.3.3 class III. Avoid skin and eye-contact by using gas-masks, gloves, masks etc. In cases of poisoning, work should be stopped. In serious cases artificial respiration might be necessary. Hospitalisation is rarely necessary.
Environmental effects: Not dangerous to bees. If birds eat the pesticide, they suffer from diarrhoea and lay wind-eggs. No cumulative effects in tissue or in the food chains.
Registration: In the US the pesticides have been placed on the list of suspicious pesticides and in other countries also, permission to use the pesticides is under pressure. However, it has never yet been prohibited.
List of some trade names, and the common names of the active ingredients as used in Appendix I.

Trade name	Active ingredient	Trade name	Active ingredient
AAtrex	Atrazine	Atred	Atrazine
Acaraben	Chlorobenzilate	Atrozol	Atrazine
Accothion	Fenitrothion	Azodrin	Monocrotophos
Actazin	Atrazine	Belt	Chlordane
Agrifume	Ethylene dibromide	Benfos	Dichlorvos
Agroxone	MCPA	Benlate	Benomyl
Akar	Chlorobenzilate	Benzilan	Chlorobenzilate
Aktikon	Atrazine	Beosit	Endosulfan
Alltox	Camphechlor	BHC	Lindane
Altan	Captan	Bibisol	Dichlorvos
Anatox	Camphechlor	Bioallethrin	Pyrethrines
Antie-carie	Hexachlorobenzene	Bladan	Parathion
Antinonnin	DNOC	Bromofume	Ethylene dibromide
Argesin	Atrazine	Brush Rhap	2,4,5-T
Atazinax	Atrazine	Brush killer	2,4,5-T
Atranex	Atrazine	BSI	2,4-D

Trade name	Active ingredient	Trade name	Active ingredient
Bunt-no-more	Hexachlorobenzene	Erasekt	Dichlorvos
Canogard	Dichlorvos	Esgram	Paraquat
Captane	Captan	Fenvalarate	Pyrethrines
Carbofos	Malathion	Flit 406	Captan
Celmide	Ethylene dibromide	Folbex	Chlorobenzilate
Chemphene	Camphechlor	Folidol	Parathion
Cleansweep	Diquat	Folithion	Fenitrothion
Cleansweep	Paraquat	Fosferno	Parathion
Clobex	Chlorobenzilate	Frumin AL	Disulfoton
Clor-chem T-590	Camphechlor	Furadan	Carbofuran
Coopervap	Dichlorvos	Gammexane	Lindane
Copper-	see Copper compounds	Gesapon	DDT
Curaterr	Carbofuran	Gesaprim	Atrazine
Cyclodan	Endosulfan	Gesarol	DDT
Cyfen	Fenitrothion	Gramonol	Paraquat
Cygon	Dimethoate	Gramoxone	Paraquat
Cypermethrin	Pyrethrines	Gramuron	Paraquat
Cytel	Fenitrothion	Granulox	Disulfoton
Cythion	Malathion	HCB	Hexachlorobenzene
Daphene	Dimethoate	HCB	Hexachlorobenzene
Dedetane	DDT	HCH	Lindane
Dedevap	Dichlorvos	Hedonal M	MCPA
Deltamethrin	Pyrethrines	Hexa-CB	Hexachlorobenzene
Denapon	Carbaryl	Hostathion	Triazophos
Detmol	Chlorpyrifos	ISO	2,4-D
Devigon	Dimethoate	Iso-cornox	MCPA
Devipan	Dichlorvos	Kopmite	Chlorobenzilate
Dextrone X	Paraquat	Linate	Zineb
Dexuron	Paraquat	Lonacol	Zineb
Dicarbam	Carbaryl	Lorsban	Chlorpyrifos
Didimac	DDT	Loxiran	Chlorpyrifos
Dieldrex	Dieldrin	Mafu	Dichlorvos
Dieldrite	Dieldrin	Malathon	Malathion
Dimetate	Dimethoate	Maldison	Malathion
Dimilin	Diflubenzuron	Malix	Endosulfan
Disyston	Disulfoton	Manzate	Maneb
Dithane M-45	Mancozeb	Manzate 200	Mancozeb
Dithane M-22	Maneb	Manzeb	Mancozeb
Dithane-Z-78	Zineb	Marvex	Dichlorvos
Dithiosystox	Disulfoton	Mecoprop	MCPP
Dowfume 85	Ethylene dibromide	Mercaptation	Malathion
Dowicide G	Pentachlorophenol	Merpan	Captan
Dowicide EC7	Pentachlorophenol	Murvin	Carbaryl
Dursban	Chlorpyrifos	Mutox	Dichlorvos
Dwco 197	Chlorpyrifos	Nemafume	Ethylene dibromide
Dyvos	Dichlorvos	Nendrin	Endrin
Edabrom	Ethylene dibromide	Neocid	DDT
Edil CP	Chlorpyrifos	NIP	Nitrofen
Ekatin-TD	Disulfoton	Niran	Parathion
Elancolan	Trifluralin	Nitrador	DNOC

Trade name	Active ingredient	Trade name	Active ingredient
Nogos	Dichlorvos	Solvirex	Disulfoton
Nutrax	Dichlorvos	Sumithion	Fenitrothion
Nuvacron	Monocrotophos	Temik	Aldicarb
Nuvan	Dichlorvos	Temizid	Aldicarb
Octachlor	Chlordane	Tersan 1991	Benomyl
Octalene	Aldrin	Tetramethrin	Pyrethrines
Octalox	Dieldrin	Thifor	Endosulfan
Orthocide 406	Captan	Thiodan	Endosulfan
Orthocide	Captan	Thiophos	Parathion
Para-col	Paraquat	Tiezene	Zineb
Parsolin	Disulfoton	Tok E-25	Nitrofen
Pathclear	Diquat	Tokkorn	Nitrofen
Pathclear	Paraquat	Tota-col	Paraquat
Patrin	Carbaryl	Toxakil	Camphechlor
PCP	Pentachlorophenol	Toxaphene	Camphechlor
Penta	Pentachlorophenol	Transamine	2,4,5-T
Perfekthion	Dimethoate	Treflan	Trifluralin
Permethrin	Pyrethrines	Trifina	DNOC
Pestmaster	Ethylene dibromide	Triflurex	Trifluralin
Phosvit	Dichlorvos	Trifocide	DNOC
Pomite	Chlorobenzilate	Trimegol-50	Captan
Primatol	Atrazine	Trimetion	Dimethoate
Pyrinex	Chlorpyrifos	Trioxone	2,4,5-T
Ravyon	Carbaryl	Trizilin	Nitrofen
Rebelate	Dimethoate	Tumbleweed	Glyphosate
Reglone	Diquat	UC 21149	Aldicarb
Rogor	Dimethoate	Vancide 89	Captan
Roundup	Glyphosate	Vapona	Dichlorvos
Roxion	Dimethoate	Vectal	Atrazine
Roxo	Dichlorvos	Velsicol 104	Heptachlor
Sandozebe	Mancozeb.	Vondocaptan	Captan
Sanocide	Hexachlorobenzene	Weedone	MCPA
Santobrite	Pentachlorophenol	Weedone	2,4,5-T
Selinon	DNOC	WSSA	2,4-D
Sevin	Carbaryl	Yaltox	Carbofuran
Sinox	DNOC	Zeazin	Atrazine
Solvigran	Disulfoton	Zidil	Chlorpyrifos

Appendix II
Annotated Bibliography

This list of publications for a great deal is taken over from A catalogue of IPM training and extension material by F.A.N. van Alebeek. This book contains a lot of descriptions of handbooks, field manuals, pocket guides, brochures, slide sets, posters, films and videos, a directory of IPM research and information, adresses of research institutes, international information centres and book shops and organizations active in IPM or related subjects. The book is available from CTA (see appendix III, under The Netherlands) and recommended for anybody working in pest management.

PEST MANAGEMENT:

Principaux ennemis des cultures de la region des grands lacs d'Afrique Centrale.
A. Autrique, 1981 (rev. ed., 1989) 144 p., 129 coloured illustr., French.
Provides descriptions and photographs of the major insect pests and fungal, bacterial and viral diseases of 20 crops in Central Africa. Notes on the life cycle, symptoms, damage and cultural and chemical control measures are included. Precautions for safe handling of pesticides are also given. Suitable for farmers, and extension workers in Central Africa.
Available from Institut des Sciences Agronomiques du Burundi (ISABU), B.P. 795, Bujumbura, Burundi.

Plant protection: a summary of principles, methods and products, for use in agricultural training in tropical and subtropical regions.
The Hague, The Netherlands, Ministry of Agriculture and Fisheries, 1986, 45 p., English.
A guide to training in crop protection providing general, basic principles and methods. Chapters deal with crop injury, prevention of injury, direct control of biotic factors, chemical control, composition of pesticides, forms of pesticide application and methods, application equipment, classification of pesticides, fungicides, insecticides acaricides, nematicides, rodenticides, safety precautions, weeds and weed control, and the application and classification of herbicides. Suitable for plant protection staff, teachers, and extension workers in the tropics and subtropics.
Available from TOOL (see appendix III) for Dfl.8.00, excluding postage.

Pest management.
G.A. Matthews, Harlow, UK, Longman Group, 1984, 231 p., 76 black & white illustr., English, ISBN 0-582-47011-0.
Covers ways of assessing crop losses, assessing pest populations (sampling and traps), methods of pest control (cultural, biological, chemical, integrated control, use of behaviour-modifying chemicals), determination of pesticide dosages, application methods and techniques, evaluation of pesticide effectiveness and effects on non-target organisms, and the concept of pest management.
Available from most bookshops.

Tropical plant diseases.
H. David Thurston, 1984, 208 p., 70 black & white illustr. English, ISBN 0-89054-063-2.
Covers 14 important tropical crop groups. A brief introduction to each specific crop or group of crops is followed by a discussion of their major diseases, and references for identification and control of pathogens. The differences between climates, soils, and farming systems are described, as well as social, political and economic factors relating to tropical crops and those grown in temperate regions. A chapter is included on international agencies concerned with tropical plant diseases.
Available from APS (3340 Pilot Knob Road, St. Paul, Minnesota 55121, USA) for US $20.00.

BIOLOGICAL CONTROL:

Organic pest control: preliminary database for program formulation.
SIBAT, 1987, 102 p., 27 black & white illustr., English.
Provides information on pest control without chemical pesticides. Chapters deal with bio-sprays (the use of plant extracts and other mixtures), biological control measures, and cultural and physical methods of crop protection. Appendices provide local names of common insect pests in the Philippine provinces, information on important natural enemies, lists of major pests of rice, corn, vegetables and legumes, and the side effects of pesticides. Suitable for farmers, and extension workers in Asia.
Available from SIBAT (Spring of science and appropriate technology, POBox 375, Manila, Philippines).

Theory and practice of biological control.
C.B. Huffaker & P.S. Mesenger (eds.), New York, USA; London, UK, Academic Press, 1976, 788 p., 40 black 7 white illustr., English, ISBN 0-12-360350-1.
Basic handbook on biological control. This book is divided into sections dealing with history and ecological basis of biological control; biological control in specific problem areas including tropical crops and fruits; elements of integrated control including biological control, selective pesticides, cultural control, host plant resistance, genetic control, and the integration of methods.
Available from most bookshops.

CHEMICAL CONTROL:

Formulating pyrethrum.
Pyrethrum Bureau, 1987, 3rd ed., 56 p., coloured illustr., English.
Describes the active consequences of pyrethrum, its biological, physical and chemical characteristics, formulated products, and the formulation theory. Appendices include information for the standard value required, proper formulation, and cured pyrethrum formulation. Suitable for farmers, extension workers, pest control operators and agricultural students.
Available from Pyrethrum Bureau, P.O.Box 420, Nakuru, Kenya, free for charge apart from postage and handling.

Pest control safe for bees: a manual and directory for the tropics and subtropics.
M. Adey, P. Walker & P.T. Walker, 1986, 225 p., 121 black & white illustr., English, ISBN 0-86098-184-3.
This handbook is divided into a manual and a directory. The manual gives general information on bee-keeping and pest control methods which are safe for bees. The directory is a comprehensive pest control manual for 85 important crops and their major insect pests. Cultural measures are recommended and chemical control methods are mentioned, if they are not harmful to bees. Suitable for farmers, extension workers, pest control operators and agricultural students in the tropics and subtropics.
Available from International Bee Research Association, Hill House, Gerrard Cross, Bucks SL9 ONR, UK.

Fundamentals of pesticides: a self-instruction guide.
G.W. Ware, Fresno, USA, Thompson Publications, 1986, 2nd rev. ed., 274 p., English, ISBN 0-913702-35-8.
A self instruction guide including a general introduction to pesticide chemistry and formulation, and detailed discussions of the chemical groups of pesticides such as herbicides, fungicides, bactericides, nematicides, rodenticides, plant growth regulators, and defoliants. The guide is comprised of 15 units, each concluding with progress checks in the form of questions. Although oriented on situations in North America, it is also useful for other geographical regions.
Available from most bookshops, price unknown.

The pesticide manual: a world compendium.
C.R. Worthing & S.B. Walker (eds.), Thornton Health, UK, British Crop Protection Council, 1987, 8th ed., 1081 p., English, ISBN 0-948404-01-9.
A worldwide reference guide to pesticide and their characteristics which compiles information about some 550 pesticides. Each pesticide is listed under its common name, giving chemical structure, nomenclature, physical and chemical properties, principal uses, toxicological data, formulations and analysis methods.
Available from most bookshops, price unknown.

Pesticide application methods.
G.A. Matthews, London, UK, Longman Group, 1979, 334 p., approx. 200 black & white illustr., English, ISBN 0-582-46054-9.
Practical and technical information on pesticide application, illustrated with numerous photographs and drawings. Following an introduction on chemical control, chapters deal with formulations, droplets, nozzles, hand-operated hydraulic sprayers, air-carrier sprayers, controlled droplets application, fogging, dust application, aerial application, injection and fumigation, maintenance of equipment, safety precautions, and the selection of equipment. Suitable for field workers and pest control operators.
Available from GTZ (Postfach 5180, D-6236 Eschborn, Federal Republik of Germany).

Manual on surveillance and early-warning techniques for plant-pest and diseases.
J. Schafer, 1982, 228 p., English.
Presents standardized procedures for use in an efficient plant protection service, in the form of guidelines for a surveillance and early-warning system. The coding system for report is described, followed by detailed descriptions of all pests (insects, weeds, rodents, birds, nematodes) and diseases encountered, and advice for their control. Suitable for extension workers and pest control operators.
Available from GTZ (see above).

Natural crop protection: based on local farm resources in the tropics and subtropics.
G. Stoll, 1987, 2nd ed., 188 p., 96 black & white illustr., English (also available in Spanish, German and French), ISBN 3-924333-43-2.
Following an introduction to the principles of preventive crop protection, insect pests of rice, maize, legumes, vegetables, fruits and stored products are described. Host plants, distribution, life cycle, damage pattern and possible control measures are discussed. The main part of the book then describes crop and storage protection methods, including the use of insecticidal plants, ashes and other substances, baits and traps. Suitable for large and small scale farmers, and extension workers in the tropics.
Available from Verlag Josef Margraff (Raffeisenstrasse 24, D-6070 Langen, Federal Republik of Germany) for US $15.00 or DM 25.00, excluding postage and handling.

STORAGE METHODS:

Small farm grain storage: preparing grain for storage (vol.1); enemies of stored grains (vol.2); storage methods (vol.3).
C. Lindblad & L. Druben, 1977, 203 p., 169p. and 147 p. respectively, 50 black & white illustr., English.
Set of handbooks covering all aspects of grain storage. Volume 2 comprehensively deals with the pests encountered in storage, and basic methods of protection. Technical information on silo construction and hygiene is provided, together with practical advice on pest control, with and without pesticides. Suitable for farmers and extension workers.
Available from VITA (1815 North Lynn St., Suite 200, Arlington VA 22209, USA); or TOOL (see appendix III) for Dfl.17.00 per volume, excluding postage.

PESTS, DISEASES, WEEDS:

Locust handbook.
A. Steedman (ed.), 1988, 2nd ed., 187 p., 152 black & white illustr. and 4 coloured plates, English, ISBN 0-85954-232-7.
Compiles basic information on locusts, their distribution, damage, life cycle and behaviour, seasonal movements and swarming. Some 20 African and South-East Asian locust species are described in detail, and methods for locust control (including a chapter on natural control) are dealt with. Suitable for plant protection officers, pest control operators and extension workers engaged in locust control in Africa.
Available from ODNRI (College house, Wrights Lane, London W8 5SJ, Great Britain) for £24.00, including postage and handling.

Insect pest control, with special reference to African culture.
R. Kumar, London, UK, Edward Arnold Publishers, 1984, 298 p., 27 black & white illustr., English, ISBN 0-7131-8083-8.
A basic textbook on insect pest control. Information is provided about insect pests, crop loss assessment and economic thresholds, forecasting and monitoring, and control methods including physi-

cal, cultural, biological, genetic and chemical control. Use of host plant resistance, formulation and application of pesticides, behaviour-modifying chemicals and aspects of pest management are discussed. Suitable for agricultural students, trainers, extension officers, and all other interested readers in Africa.
Available from most bookshops.

Agricultural insect pests of the tropics and their control.
Dennis S. Hill, Cambridge, UK, Cambridge University Press, 1983, 2nd ed., 746 p., approx. 700 black & white illustr., English, ISBN 0-521-24638-5.
A general introduction to insect pests in the tropics. Chapters deal; with the principles of pest control, methods for pest control, pest damage, biological control and chemical control. The book comprises descriptions of some 300 insect and mite pests and gives overviews of pest spectra in about 100 crops. Suitable for extension workers, pest control operators, agricultural students and scientists in the tropics.
Available from most bookshops.

Les nematodes parasites des cultures maraîchères: introduction aux nematodes phytoparasites.
J.C. Prot, ORSTOM, 28 p., 13 black & white and 22 coloured illustr., French (also available in English).
A general introduction to nematology, including morphology, biology, reproduction, and relationship to host plants and the environment. Special attention is paid to the Meloidogynae in Senegal vegetables, their development, symptoms, and control methods. Suitable for extension workers, pest control operators and agricultural students.
Available from USAID Regional Food Crop Protection Project 6250928, Dakar, B.P. 49, Sénégal.

Diseases, pests and weeds in tropical crops.
J. Kranz, H. Schmutter & W. Koch (eds.), Berlin and Hamburg, Federal Republic of Germany, Verlag Paul Parrey, 1977, 666 p., 238 black & white illustr., and 250 coloured photographs, English (also available in German, French and Spanish), ISBN 3-489-68626-8.
A recommended guide to crop protection in the tropics, dealing comprehensively with all major pests, diseases and weeds. The emphasis is on symptoms, biology, ecology and control of organisms harmful

to important tropical field crops. Chapters deal with diseases, pests and weeds, organizing each into taxonomical groups. Pests, weeds or diseases in specific crops can be selected via the indexes.
Available from most bookshops.

Virus diseases of important food crops in tropical Africa.
H.W. Rossel & G. Thottappilly, 1985, 61 p., English. Information about the geographical distribution, symptoms, identification and control of the most prevalent virus diseases of some of the continent's principal staple food crops. Suitable for extension workers and agricultural students in Africa.
Available from International Institute of Tropical Agriculture (IITA, P.M.B. 5320, Ibadan, Nigeria) for US $5.00, excluding postage and handling; add $12.00 for airmail and $7.00 for surface mail.

Weed science in the tropics: principles and practices.
E. Sauerborn & J. Sauerborn, Universitat Hohenheim, Federal Republic of Germany, Institut für Pflanzenproduktion in den Tropen und Subtropen, 1988, 264 p., approx. 150 black & white illustr., English, ISBN 3-924333-55-6.
Describes some 300 species of weeds, including details about their distribution, habitat, morphology, synonyms and local names.
Available from Jozef Margraff Verlag (see above) for approx. US $25.00.

The biology of weeds.
T.A. Hill, London, UK, Edward Arnold, 1977, Studies in Biology no.79, 64 p., black & white illustr., English, ISBN 0-7131-2637-X.
A comprehensive and easy-to-read introduction to weeds and their biology. Suitable for student textbook.
Available from most bookshops.

Rodent pests and their control.
N. Weis, 1981, GTZ Special Issue, 200 p., 70 black & white and coloured illustr., English.
Destructive rodent species, their biology, and preservation instructions for collectors are first described, together with guidelines for planning a campaign and evaluating its results. Later chapters present methods for controlling rodent pests in built-up areas and open country, and advise on use of, and

resistance to, rodenticides. The correct application of bait and construction of bait containers is illustrated, and the use of proven appraisal methods in sugar cane, rice and maize is demonstrated. Accident prevention and first aid measures are also discussed. The rodent pest situation in various countries is assessed. Suitable for extension workers. Available from GTZ (see above).

SAFE USE OF PESTICIDES:

Guidelines for the safe and effective use of pesticides.
GIFAP, 1983, 58 p., 50 coloured illustr., English (also available in Spanish, French and Portuguese). Clearly illustrated, useful brochure dealing with all stages of pesticide use, from initial selection of a pesticide to application in the field. Guidelines are given for safety precautions, avoidance of risks, and handling in case of accidents at each stage. First aid in case of pesticide poisoning is briefly described. An accompanying pictorial poster and series of slides are also available. Suitable for farmers in general, extension workers, pest control operators and agricultural students.
Available from GIFAP (see appendix III) for BF150.00 if less then 35 copies are ordered, quoting GSUBE (GSUBS for the Spanish, GSUBF for the French, or GSUBP for the Portuguese version).

Toxicology and safe handling of pesticides.
Philippine-German Crop Protection Programme, 1987, 56 p., 19 black & white illustr., English. Brochure dealing with all major aspects of safe pesticide use. Information is provided on toxicology and hazards of pesticides, formulations, pesticide labels, purchase, storage and application of pesticides, calibration of sprayers, pesticide calculations, maintenance of application equipment, residues, and disposal of pesticide solutions and containers. Appendices recommend first aid and treatment regimes in case of pesticide poisoning. Suitable for farmers, pest control operators, plant protection officers and extension workers.
Available from Philippine-German Crop Protection Programme, Bureau of Plant Industry, Ministry of Agriculture, San Andres, Malate Manila, Philippines.

Knapsack sprayers: use, maintenance, accessories.
F. Fraser & L.C. Burrill, 1979, 31 p., 75 illustr., English.
Manual describing functions and details for constructing a variety of multi-nozzle booms for use in conjunction with manually-operated hydraulic sprayers. Sprayer operation, maintenance, and calibration are also discussed. Suitable for pest control operators, plant protection officers and extension workers.
Available from IPPC (Oregon State University, Corvallis, Oregon 97331, USA) for US $3.00, including surface mail.

EXPERIMENTING:

How to perform an agricultural experiment.
G.S. Pettygrove, 26 p., English ISBN 0-86619-039-2 (also available in Spanish, ISBN 0-86619-040-6). Brochure describing the basic considerations for the design, execution, and measurements procedures of an agricultural experiment. Suitable for agricultural students and extension workers.
Available from VITA Publication Services (see above) for US $7.50 surface mail, or US $9,75 airmail outside the USA.

VARIOUS CROPS:

Illustrated guide to integrated pest management in rice in Tropical Asia.
W.H. Reissig et al., 1986, 411 p., over 1000 black and white illustr., list of terms, English, ISBN 971-104-120-0.
Systematically covers all aspects of pest management in rice, explaining each topic step-by-step with drawings. Chapters deal with the rice plant, about 40 species of insect pests, 16 kinds of diseases, 15 weed species, rats, cultural control, resistant varieties, biological control, pesticides, integration of control measures, and implementation of IPM-strategies. Suitable for farmers, extension workers, pest control operators, and agricultural students in Tropical Africa.
Available from IRRI (POBox 933, Manila, Philippines).

Compendium of ... diseases.

Approx. 125 p., with coloured and black and white illustr., English.

A series of compendia, which describes major diseases of many crops. Following a general introduction to the crop, chapters deal with fungal diseases, bacterial and bacteria-like diseases, nematode parasites, nutritional deficiencies, toxicities and other abiotic processes, insect pests and the control of diseases. Suitable for farmers, pest control operators, agricultural students, and extension workers. Written for the USA, although it may also be relevant to other geographical regions.

Compendia are available for:

barley (1982, ISBN 0-89054-047-0),
bean (1988, in preparation),
citrus (1988, in preparation),
corn (1980, ISBN 0-89054-029-2),
cotton (1981, ISBN 0-89054-031-4),
pea (1984, ISBN 0-89054-060-8),
peanut (1984, ISBN 0-89054-055-1),
potato (1981, ISBN 0-89054-027-6), also a Spanish version,
sweet potato (1988, ISBN 0-89054-089-6),
sorghum (1986, ISBN 0-89054-069-1),
soybean (1982, ISBN 0-89054-043-8),
wheat (1987, ISBN 0-89054-076-4),

Available from APS Press (3340 Pilot Knob Road, St. Paul, Minnesota 55121, USA) for US$ 25.00, including postage.

Pest control in tropical root crops.

Centre for Overseas Pest Research, PANS Manual no. 4, black & white illustr., English.

A complete description of weeds, diseases, nematodes, insect and mite pests of tropical root crops. Distribution, transmission, and control methods (mainly cultural and chemical control) are given. Suitable for pest control operators and extension workers.

Available from ODNRI (see above) for £3.25.

Control integrado de plagas de papa.

Luis Valencia, CIP, 1986, 203 p., 84 black & white and coloured illustr., Spanish.

Collection of 20 articles used in the 1986 course on integrated pest control for potatoes, by the International Centre for Potato Research in Colombia. The articles deal with chemical and biological means of pest control, descriptions of important groups of pests, and current practices in Colombia concerning integrated pest control in potatoes. Suitable for extension workers, pest control operators and agricultural students in Colombia and Latin-America.

Cassava pests and their control.

A. Bellotti & A. van Schoonhoven, 1978, 73 p., 60 coloured illustr., English.

Discusses host plants, pest distribution, mites and insects attacking the crop, and economic damage. A recommended integrated cassava pest management programme is described. Suitable for farmers, extension workers and pest control operators.

Available from CIAT (see appendix III).

Yuca; control integrado de plagas.

J.A. Reyes (compiler), 1983, 362 p., Spanish.

Reference book for CIAT training courses on integrated control of cassava pests. Contains chapters on the principles of integrated control, crop losses due to insect and mite damage, descriptions and biology, the use of resistant varieties, biological control and integrated control. Suitable for extension workers, pest control operators, and agricultural students.

Available from CIAT (see appendix III).

Guia de control integrado de plagas en maiz y sorgo.

Alvaro Sequeira D. et al., 1979, 44 p., 17 black & white illustr., Spanish.

This pocket guide introduces several agricultural practices which can be used as part of an IPM system, then briefly describes 16 pest species of maize and sorghum. Notes on their importance, damage they cause, and control are included. Appendices give more information on available insecticides, their toxicity, and safety precautions. Suitable for farmers, extension workers, and pest control operators in Latin America.

Available from Instituto Nicaraguense de Tecnolia Agropecuaria (INTA), Proyecto de Control Integrado de Plagas, Managua, Nicaragua.

Insect pests of maize: a guide for field identification.

C. Alejandro Ortega, CIMMYT, 1987, 106 p., 96 coloured illustr., English (also available in Spanish and French), ISBN 968-6127-07-0.

Approximately 70 species or genera of insect pests

of maize are shown in coloured photographs, and the nature of their damage, life cycle and geographic distribution is described. A simplified key identifies the pest according to the growth stage of the maize plant and the position of the pest organisms on the plant. An introduction to insect pests, beneficial insects in maize, and pest control (economic thresholds) is also included. Suitable for farmers, pest control operators and scientists in the tropics and subtropics.

Available from CIMMYT for US $7.50 (HDC), or US $5.50 (LDC).

Integrated pest management: corn, a pocket reference manual.

Philippine-German Crop Protection Programme, 1987, 109 p., 46 coloured illustr., English, ISBN 971-91057-1-2.

A comprehensive introduction to IPM for corn in the Philippines. Symptoms or damage characteristics, alternative hosts, affected growth stages, and pest management recommendations (cultural methods, biological control, pest monitoring, economic threshold levels, chemical control and resistant varieties) are described for 10 insect pests, 10 diseases, weeds and rodents. Physiological disorders are illustrated and described, and a separate section details several IPM prevention and control recommendations, especially biological pest control. A diagnostic guide for corn diseases in the Philippines is provided, as well as growth stages for corn. Suitable for farmers and extension workers in Asia.

Available from Philippine-German Plant Protection Programme, Bureau of Plant Industry, Ministry of Agriculture and Food, San Andres, Malate, Manila, Philippines.

Sorghum insect identification handbook.

L. Teetes et al., ICRISAT, 1983, Information Bulletin no.12, 124 p., 60 coloured illustr., English (also available in Spanish and French).

Describes the common insect and mite pests of sorghum which attack the roots, foliage, stem or head, including their distribution, symptoms, and biology. A number of important natural enemies are also described, as well as pests of stored sorghum grains. Notes on cultural control methods, economic threshold levels (if known) and chemical control methods are given for each pest. Suitable for

extension workers, pest control operators and agricultural students.

Available from ICRISAT (Patancheru P.O. Andhra Pradesh 502 324, India) for US $8.10 (HDC) or US $2.70 (LDC); add approximately $2.70 for postage.

Manual d'identification des maladies du sorgho et du mil.

R.J. Williams, R.A. Frederiksen & J.C. Girad, ICRISAT, 1978, Bulletin d'Information no.2, 88p., 59 coloured illustr., French (also available in English and Spanish).

A pocket guide briefly describing diseases caused by fungi, viruses and bacteria, which uses coloured photographs to illustrate symptoms. Information on the life cycle is given, and keys aid identification of the diseases in sorghum and millet. Suitable for farmers, extension workers, pest control operators and agricultural students.

Available from ICRISAT (see above).

Common diseases of small grain cereals: a guide to identification.

F.J. Zilinsky, 1983, 150 p., 350 coloured illustr., English (also available in Spanish).

A general introduction to the identification of common fungal, bacterial and viral diseases of grain cereals. Symptoms are shown in coloured photographs, accompanied by short descriptions of the diseases, their biology, importance, and general remarks on control methods. Suitable for extension workers, pest control operators and agricultural students.

Available from CIMMYT (see appendix III) for US $20.00 (HDC) or US $12.00 (LDC), excluding postage and handling.

Pest control in tropical grain legumes.

Centre for Overseas Pest Research, approx. 250 p., more than 100 black & white illustr., English.

One of a series of detailed manuals about pest control in specific crops. The general agronomy of the crop, weeds and their control, diseases, nematodes, invertebrate and vertebrate pests, and insect pests are dealt with. Suitable for extension workers and pest control operators. From the same series: Pest control in tropical **onions**, Pest control in **Bananas**, Pest control in tropical **tomatoes**.

Available from ODNRI (see above) for £5.25.

Cowpea production training manual.
IITA, 1982, 198 p., English (also available in French).
A basic reference manual for IITA's cowpea training course for the humid and sub-humid tropics. History, origin, importance, distribution, botany, physiology, agronomy, pathology, entomology, and breeding are included, as well as seed production and distribution. Suitable for extension workers in humid and sub-humid tropics.
Available from IITA, free of charge.

Plagas e enfermedades de algodon en Centro America.
H. Schmutterer, 1977, Schriftenreihe der GTZ no.39, 104 p., 50 coloured illustr., Spanish.
An advisory brochure about the integrated control of cotton pests in Central America. Suitable for farmers, extension workers and pest control operators.
Available from GTZ (see above) for DM 22.00

Appendix III
List of organizations

This appendix lists some organizations which are actively involved in IPM, safe or more adequate pesticide use, or traditional and modern agriculture.
The list is ordered alphabetically by country. The fatprinted code names give an indication of the working field of the organization.

safe use   organization working on safe and adequate use of pesticides;
pois   organization working on the problems regarding pesticide poisoning;
ipm   organization active in the field of IPM (integrated pest management);
tradagr   organization working on traditional agriculture as alternative to conventional agriculture;
modagr   organization occupied by modern, alternative agricultural methods as alternative to conventional agriculture;
env   organization occupied by environmental problems in general.

The Pesticide Action Network (PAN, see chapter 16.1.1).
The five regional PAN centres, of which the european is the central one, are:

Africa   Environment Liaison Centre (ELC), PO Box 72461, Nairobi, Kenya.
Asia and the Pacific   IOCU, PO Box 1045, Penang, Malaysia.
Europe   ICDA, Damrak 83, 1012 LN Amsterdam, The Netherlands.
Latin America   Fundación Natura, Casilla 243, Quito, Ecuador
North America   Friends of the Earth, 1045 Sansome St., San Francisco, CA 941211 U.S.A.

Australia

Toxic and Hazardous Chemicals Committee (safe use)
Total Environment Centre,
18 Argyle Street,
Sydney 2000,
This is an action information centre which campaigns against unsustainable development and the misuse of resources. It sponsors educational workshops, conducts independent research and sponsors several standing committees, one of which deals with toxic and hazardous chemicals. This committee has been very active in challenging the overuse of pesticides in Australia and has promoted alternatives.

Belgium

GIFAP (pois)
Avenue Hamoir 12
1180 Bruxelles
GIFAP is an international association of national chemical producers. It publishes the monthly GIFAP bulletin which discusses pesticides from the industrial point of view, covers toxicology data and upcoming agrochemical meetings. It has also produced a series of how-to booklets in English, French, Spanish and Portuguese. Copies are available on request.

Friends of the Earth (env)
25, rue du Prince Royal
B-1050 Bruxelles

Stichting Leefmilieu (env)
Keizerstraat 8
B-2000 Antwerpen

Brazil

Acao Democratica Feminina Gaucha (ADFG) (pois)
Rua Lucas de Oliveira
1250-90000 Porto Alegre
The ADFG is at present mounting a campaign for national regulatory controls. Previous actions on the state level had initial positive results but regulatory law propositions were overruled in the national Supreme Court enforced by an industrial lobby.

Os Amigos da Terra (env)
(Friends of the Earth)

Rua Botucatu, 30
04023 Sao Paulo, SP

Canada

The Developing Countries Farm Radio Network (DCFRN) (safe use)
c/o Massey-Ferguson Ltd.
595 Bay Street
Ontario M5G 2C3
This NGO collects and disseminates information on practical and simple techniques to increase Third World food production and to improve nutritional and health status. The network, based in Canada, produces scripts in English, French and Spanish, in written and tape cassette form which are sent to various parts of the world to be played on local radio stations.

Friends of the Earth (env)
53-53 Queen Street
Ottawa, Ontario K1P 5C5

Colombia

Centro Internacianal de Agricultura Tropical (CIAT)
Apartado Aereo
6713 Cali

Ecuador

Fundación Natura (safe use/pois)
Jorge Juan 481 y Mariana de Jesus
Quito
This NGO is a founding member and regional focal point of PAN (see above) and has been quite active in the pesticide issue at national and international level.

Federal Republic of Germany

Deutsche Gesellschaft für Technische Zusammenarbeit
Dag-Hammarskjöld-weg 1+2,
Postfach 5180
D-6236 Eschborn

France

International Federation of Organic Agriculture Movements (IFOAM) (modagr)
Chateau de Chamarande
91730 Chamarande
This international federation consists of 80 member

groups from 30 countries and represents about 80,000 individuals. Membership is open to primary producer groups and to agricultural research, development, education and media groups who agree with the aims of IFOAM. The aims are to promote an ecologically, economically and socially sustainable agriculture. The federation publishes three bulletins in French, Spanish, and English which provide information of both scientific and practical importance. It also sponsors bi-annual scientific conferences. Proceedings from previously held meetings can be obtained.

Great Britain
OXFAM (pois/ipm)
274 Banbury Road,
Oxford OX2 7DZ
This development agency is a founding member of PAN. The Public Affairs office conducts research on pesticide use in the Third World.

Friends of the Earth Trust Limited (env)
377 City Road
London, EC1V 1Na

India
Association for the Propagation of Indigenous Genetic Resources (tradagr/modagr)
Naroda Road
Ahmedabad 380025
This organization dedicates itself to organic farming and the preservation of indigenous cereal seed and tree germ-plasm.

Ghandi Peace Foundation (env)
221/3 Deen Dayal Upadhaya Marg
New Delhi 110002
This NGO promotes environment oriented development through rural development institutions and works with other organizations on forestry projects.

Indonesia
Centre for Environmental Studies (PKLH) (env)
Jalan Prof. Maas No. 3A
Kampus USU
Medan
North Sumatra

Wahana Lingkungan Hidup Indonesia (WALHI) (env)
(Indonesian Environmental Forum)
Jalan Suryopranoto 8
Lantai IV
Jakarta Pusat
WALHI is concerned with the range of environmental problems created by rapid and widespread development, including problems of overpopulation, destruction of forests, fauna and marine life, and pollution.

Ireland
Friends of Earth (env)
26 Millimount
Drumcondra
Dublin 9

Italy
Food and Agricultural Organization of the U.N. (FAO) (safe use/ipm)
Via delle Terme di Caracalla
00100 Rome
This UN organization, particularly the Department of Plant Production and Protection has done much research and published a lot of scientific and practical literature on safe use of pesticides and IPM. The FAO works with national pest control programmes in some countries in South-east Asia, and in many other countries.

Amici della Terra (env)
(Friends of the Earth)
Piazza Sforza Cesarini, 28
00186 Rome

Japan
Chikyo No Tomo (env)
(Friends of the Earth)
1-51-8, Yoyogi, Shibuya-ku
Tokyo 151

Kenya
Environment Liaison Centre (safe use/pois/ipm/tradagr/modagr)
P.O. Box 72461
Nairobi
This is an international association of over 230 NGO's based in Nairobi. It has been working on the pesticide issue for several years and a founding member of PAN and the network's regional focal point.

Mazingira Institute (safe use)
P.O. Box 14550
Nairobi
This national NGO publishes Rainbow, one of the few educational magazines specifically for children. It is a monthly published cartoon which is sent to 122,000 children in Kenya. Each year an issue is published which is sent to Ugandan and Tanzanian children.

International Council for Research (modagr)
in Agroforestry (ICRAF)
P.O. Box 30677
Nairobi
This NGO was established in 1977 as an international organization to promote and support agroforestry research and develops a Farming Systems Research methodology. It also has looked at indigenous use of wild plants as source of medicine, herbs, fodder and food.

Malaysia
International Organization of Consumer Unions
(IOCU) (safe use/pois/ipm) Regional Office for Asia and the Pacific,
P.O. Box 1045
Penang
IOCU, founded in 1960, has a membership of more than 130 consumer organizations in some 50 countries. Its concern with pesticides is of long standing as it is one of the founders as well as the Asian regional focal point for PAN.

Sahabat Alam Malaysia (env/pois)
(Friends of the Earth)
37 Lorong Birch
Penang
This NGO, member of PAN, has been very active against the overuse of pesticides and uncontrolled marketing of highly toxic brands. Their publications often carry articles on these issues.

Mexico
Pan American Centre for Human Ecology and Health (env)
P.O. Box 37-473
Mexico 6, D.F.

Centro Internacional de Mejoramiento de Maiz y Trigo (CIMMYT) (modagr)

Lisboa 27
Apartado Postal 6-641,
06600 Mexico D.F.

New Zealand
Friends of The Earth (env)
P.O. Box 39065
Auckland West

Netherlands
Technical Centre for Agricultural and Rural Co-operation (CTA) (tradagr/modagr)
POBox 380
6700 AJ Wageningen.
CTA was established in 1983 under the second Lomé convention between the member states of the European Community and 63 states from Africa, the Caribbean and the Pacific (the ACP-countries). CTA's objective is to assemble and disseminate technical and scientific information in the spheres of agricultural and rural development and extension. It initiates studies, prepares publications, organizes meetings of experts, assists local documentation centres in ACP-countries, and provides a query-response service. Besides several co-publications, CTA publishes the bimonthly bulletin "SPORE".

Seeds Action Network (SAN) (env/modagr)
Eerste Helmersstraat 106
1045 EG Amsterdam
This network was started up in march 1985 as one part of the struggle for a sustainable agriculture that will safeguard genetic resources. This network plays a significant role in the conservation of varieties of crops and wild plants with pest resistant or pesticidal characteristics.

Stichting Mondiaal Aternatief (env/pois)
P.O. Box 168
2040 AD Zandvoort
This national NGO, a member of PAN, is active at the national and international level lobbying for a more sustainable approach of agriculture and has also published reports on pesticide use world wide.

Trans National Information Exchange (TIE) (pois)
Paulus Potterstraat 20
1071 DA Amsterdam
TIE is a network of 40 organizations concerned with transnational corporations and acts as an interna-

tional forum for farmers, consumers, workers, action groups and trade unions. TIE has a task force on agribusiness which has specialized in pesticide industry.

Information Centre for Low External Input Agriculture (ILEIA) (ipm/modagr/tradagr)
P.O. Box 64
3830 AB Leusden
ILEIA was founded in 1982 by the Educational Training Consultants, a Dutch foundation and member of the ELC. The Centre's goal is to disseminate information on the exchange of experiences of those working towards a sustainable form of agriculture in which the farmer depends on local rather than imported (locally or internationally) resources. The centre produces a quarterly newsletter, ILEIA; maintains a computer data bank with over 800 references on IPM and related subjects; has a list of persons and organizations who are interested in or working on low external input agriculture; has a slides collection showing cropping systems in Brazil, Ghana, India, Rwanda and Tanzania.

Stichting Technische Ontwikkeling Ontwikkelingslanden (TOOL) (modagr/tradagt)
Entrepôtdok 68A/69A
1018 AD Amsterdam
TOOL is an interdisciplinary organization of dutch universities functioning as information and consultant agency for appropriate technology in the agricultural, technical and medical area. TOOL provides for answers on questions concerning technical problems encountered by development workers, missionaries etc.

Peru
Grupo Ecológico Natura (env)
Apartado 3051
Lima 100

Philippines
Appropriate Technology Centre (modagr)
in Rural Development (ATCRD)
P.O. Box EA-31
Ermita Manila
This NGO effectuates a training curriculum on tropical organic agriculture.

International Rice Research Institute (IRRI) (modagr)
POBox 933
Manila
Philippines

Portugal
Amigos da Terra (env)
(Friends of the Earth)
Associacao Portuguesa de Ecologistas
Pr. Ilma do Faial 14-A
1000 Lisbon

Spain
Federación de Amigos de la Tierra (env)
(Federation of Friends of the Earth)
Apartado 46.177
Madrid

Switzerland
Agrecol Development Information (tradagr/-modagr)
c/o ökozentrum
CH-4438 Langenbrück
This information centre is setting up an international focal point on sustainable agriculture and maintains a documentation centre. The centre promotes initiatives that foster long- term self-reliant agriculture favouring indigenous farming and newer alternative approaches such as organic and bio-intensive farming.

NGO Liaison Service (env)
Palais des Nations
CH-1211 Geneva 10

Thailand
Siam Environmental Club (env)
Faculty of Science
Chulalongkorn University
Bangkok
The club is established to provide informal environmental education for the public and for policy-makers, as well as to call attention to environmental problems.

Togo
Pan-African Federation of Agricultural (env/pois)
Trade Unions (PAFATU)
KO 1293 Rue Benissan Gbikpi Tokoin Wuiti

c/o B.P. 7138 Lomé
PAFATU is active on the issues of seeds, pesticides and environment ecology and works in collaboration with ELC.

United States of America

Office of Environment and Scientific Affairs (safe use)
World Bank
1818 H Street N.W.
Washington
D.C. 20433 U.S.A.
Since several years this office works on pesticide safety. They developed a set of guidelines to govern the use of pesticides in the projects funded by the Bank.

National Resource Defense Council (NRDC) (pois)
122 E. 42nd Street
New York
N.Y. 10168 U.S.A.
The NRDC is a member of PAN that has conducted research on pesticide residues in foods.

World Neighbors (safe use/env)
5116 North Portland Avenue
Oklahoma City
O.K. 73112 U.S.A.
This international NGO has been promoting north-south cooperation in rural development and is specialized in the planning, organizing and management of agricultural programs, including safe use of pesticides.

World Resources Institute (env/safe use/pois)
1735 New York Avenue N.W.
Washington
D.C. 20006 U.S.A
This policy research centre focuses on environment, development, population and resource issues.

Institute for Food and Development Policy (IFDP) (pois)
1885, Mission Street
San Francisco
CA 94103 U.S.A.
This NGO has conducted research on the general theme of global food security and addressed more specific topics such as the use of pesticides.

Ecology Action (modagr)
5798 Ridgewood Road
Willits
CA 95490 U.S.A.
Publishes Self Teaching Mini-Series and other practical literature on modern alternative agriculture.

Educational Concerns for Hunger Organization (ECHO) (modagr)
R.R.2, P.O. Box 852
North Fort Myers
Florida
33903 U.S.A.
This Christian Information Centre for subsistence food production provides training, conducts research on new farming methods, maintains a seed bank of under-exploited food plants and varieties of more common crops that will grow under adverse tropical conditions and publishes the periodical ECHO Technical Notes.

Rodale Institution (modagr)
R.D.1 Box 323
Kurtztown
PA 19830 U.S.A.
This NGO conducts research on cultural, mechanical and biological control of insect pests, on the cultivation of perennial grains and on tillage and rotation schemes. It has also conducted a long term study on the problems associated with the conversion from conventional farming practices to organic methods.

Regenerative Agriculture (modagr)
222 Main Street
Emmaus
PA 18049 U.S.A.
An association founded by the Rodale Institution, that executes a programme on food production systems that use local resources combining successful farmers' practices with the potentials discovered by formal science. The programme aims at a greater degree of self reliance in food production.

Zambia

Environmental Health Officers Association (env)
P.O. Box 735
Kabwe

Appendix IV
List of terms and abbreviations

action threshold damage threshold

acute toxicity how poisonous a pesticide is to an animal or man after a single exposure, usually expressed in LD50

biological control the control of pests employing natural means such as predators, parasites or pathogens

chemical control reduction of a pest by application of a pesticide

cultural control the use of agronomic practices such as soil tillage, varying planting time, fertility levels, sanitation, water management and short-growing cultivars to reduce pest populations

damage reduction in yield caused by pests

damage threshold the lowest population density that will cause economic damage

EC European Community

economic threshold the density of a pest population at which control measures should be carried out to prevent an increasing pest population from reaching the damage threshold

ELC Environmental Liaison Centre

FAO Food and Agricultural Organization

formulation the mixture in which a pesticide is sold for use, e.g. dust, granule, wettable powder, emulsifiable concentrate, etc.

green revolution development in agriculture, started in the sixties, by which the combined application of high yielding cultivars with artificial fertilizer and pesticides caused an enormous increase in yields, going hand in hand with enormous social problems in agricultural areas in the Third World

horizontal resistance a general resistance, controlled by many minor genes, which provides resistance (usually moderate) to all disease races or insect biotypes of a given species

host plant a plant species which serves as a source of food, shelter or a site to lay eggs for an organism

ICDA International Coalition for Development Action

IMF International Monetary Fund

injury any disformation caused by a damaging agent such as a pest, hail, frost, storm, lack or overdose of a nutrient, etc.

integrated pest management (IPM) trying to keep pests under the damage threshold by the use of combinations of two or more control methods such as biological, chemical or cultural control

IOCU International Organization of Consumer's Unions

IPM integrated pest management

IRPTC International Register of Potentially Toxic Chemicals. IRPTC registers control actions of states to ban or severely restrict a chemical and disseminates such actions to other states.

IRRI International Rice Research Institute

loss money to be lost because of damage done by a damaging agent

modern technology this term is used to indicate technologies which are developed in the West and exported from there to the Third World

NGO Non-Governmental Organization

PAN Pesticide Action Network

pathogens organism which cause diseases, for example fungi, bacteria, viruses, nematodes, etc.

pest management management of pest populations through the use of monitoring methods and the employment of control measures based on economic thresholds

pest an unwanted organism which attacks or competes with crops; in this book we use the term pest for insects, mites, rats, birds, fungi, bacteria, viruses, weeds and all other living damaging agents on crops

PIC Prior Informed Consent

prior informed consent the principle that governments of pesticide exporting countries ask impor-

ting countries for an approval of the import of a
chemical

sanitation a set of measures to ensure that as little
as possible individuals of a pest are introduced in
the field

sustainable agriculture a form of agriculture for
which natural sources are not exhausted and which
lasts for very long periods (at least 50 or 100 years)

systemic pesticide a pesticide that is absorbed by
the plant and by which pests consuming the plant
or plant sap are controlled

Third World In this publication the internationally
accepted terms "Third World" and "the West", or
"Western World" are used. There are few satisfactory alternatives to these terms, and none lend
themselves to repetitive use. We would like to em-
phasise, therefore, that the term "Third World"
should not be understood as a group of countries
which come third and last in the list of priorities
and "The West" does not indicate a group of countries situated in the western part of the world;
Japan obviously is in the East.

variety (officially: cultivar) a sub-species of a
plant, differing from other cultivars by certain characteristics

UNEP United Nations Environment Programme

UNESCO United Nations Educational and Scientific
Organization

UNIDO United Nations Industrial Development Organization

Western World see "Third World"

WHO World Health Organization

Contributors to this book

This book has been written by the Crop-protection section of the Centre for development work, The Netherlands. Many people contributed to the realization of the book. When we try to name them all, we are sure to forget some. The people we did not forget are named in the following list.

Frans van Alebeek, Rob van Alkemade, Frenk Boeren, Koen den Braber, Alfons Broeks, Marietje van Eeghen, Elske van der Fliert, Henk Geist, Fred Geven, Marion Herens, John Holland, Marijke van Hooydonk, Gerriet Kooy, Mans Lanting, Marg Leijdens, Wim de Louw, Jan Louwen, Loet van Moll, Trix Overtoom, Henk Peters, Iling Tjon Piangi, Ineke van de Pol, Frank van Schoubroeck, Huub Venne, Felix Wäckers.

Advisors involved in the realization of the book were:
James Everts, Louise Fresco, Arnold van Huis, Vera Kappers, Bert Lokhorst, Jan van de Waerdt, JaapJan van der Weel.

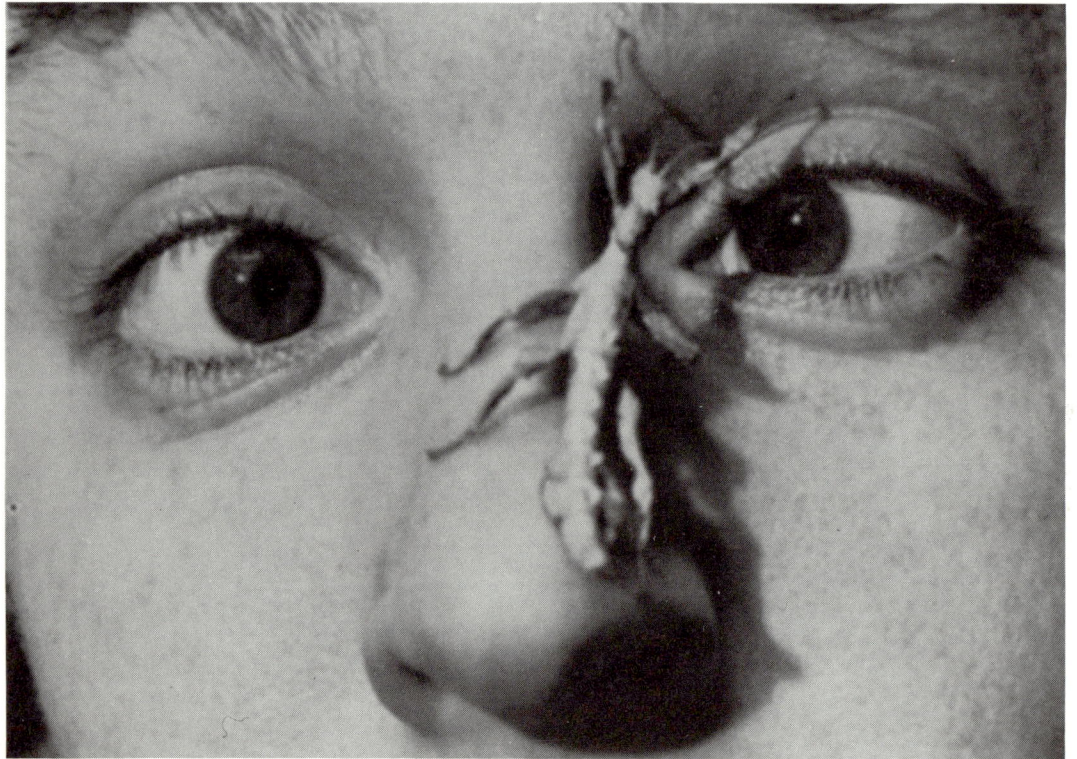